The Art of
Mathematics

The Art of Mathematics

Jerry P. King

Fawcett Columbine • New York

A Fawcett Columbine Book
Published by Ballantine Books

Contents

Introduction

Years ago we summered in Vermont. We lived way north in the kingdom of green mountains and tall trees. The dirt road passing our house had remained unchanged since the revolutionary war and was so untraveled that when a car came down we knew we had company. The days drifted past quiet and easy and the loudest sound was the occasional mooing of a cow in the pasture over the hill. Quick fish moved in the clear, cold lake and the night sky glistened with stars. Summer breezes turned the grass.

One night, with the candles burning low, the wine spent, and the moonlight just right, I said to my wife:

You are more beautiful than any woman I have ever seen.

I was looking directly at her when I spoke. She looked back. And she smiled softly. And she did not—I am happy to say—think at that moment like a mathematician.

For had she done so, she would have concluded that my compliment was nonsense and nowhere near anything that could be called truth. For, if my statement were true, then it would follow that she—being *more* beautiful than any woman I had ever seen while simultaneously being the object of my amorous gaze—would be more beautiful than herself which, of course, is impossible. Had she examined my remark with the cold light of precision she would have found it meaningless. And she would have broken the mood.

But she did not analyze. She *knew* what I meant.

And so did I. I opened another bottle of wine and blew out the candles.

Mathematicians, unlike lovers, loathe imprecision. Precision is the hallmark of mathematics and you can neither talk mathematics nor write mathematics without first committing yourself wholly to the strict adherence to rules of inference and exposition. And when you do you will capture the attention of the mathematicians. And for the proper care and handling of their subject they will also give you their appreciation. But that is all you are likely to get. If you want more, candlelight and wine will serve you better than precision.

I am writing—not *mathematics*—but *about mathematics*. And the gap between these two notions stretches wide as the distance between "physics" and "metaphysics." You write *about mathematics* for nonmathematicians. Mathematicians, like Prufrock's mermaids, sing *mathematics* only each to each. They do not sing for us.

This book is not written for mathematicians. Instead, it is aimed toward an audience of those people for whom art, music, and literature are common pursuits but for whom mathematics may stand as a deep mystery. Or as something horrid—a thing to be avoided like a nightmare.

Mathematicians know something the rest of us do not.

They know that mathematics, rightly viewed, possesses an aesthetic value as clearly defined as that of poetry. But no one else knows this. This knowledge remains locked deep inside the closed aristocracy of mathematicians.

Humanists find pleasure inside concert halls, art galleries, and great books but flee from mathematics like peasants racing from Frankenstein's monster. Meanwhile, the aesthetic experience of mathematics both excites and soothes mathematicians. The excitement brings them to the subject and the soothing becomes addictive and draws them back again and again. Their experience exactly duplicates that which time and again returns the humanist to the Venus de Milo in the Louvre or to the impressionist pictures in the Musée d'Orsay.

Mathematicians also go to galleries and concerts—not as often as they profess—but frequently. The aesthetic experiences of the humanists are open to the mathematicians. But not the other way around. The aesthetic pleasures of mathematics are unavailable to the humanists. Unavailable not because the aesthetics lie beyond their comprehension but because the *right view* of mathematics has been kept hidden from them.

The book intends to expose the view, to bring the secret out into the light. And to show that one's intellectual and aesthetic life cannot be complete unless it includes an appreciation of the power and the beauty of mathematics. Simply put, aesthetic and intellectual fulfillment requires that you know *about mathematics*.

The secrecy surrounding the aesthetics of mathematics results partly from the fact that mathematicians do not like to talk about mathematics. Neither do they like to write about mathematics. Nor to read about mathematics.

Mathematicians like to *do mathematics*. And by doing

mathematics they mean doing *research* in mathematics. And by research they mean the *creation* of new mathematics. They mean exactly this, and nothing more.

The reader must understand this distinction clearly: to write *about* mathematics is to write exposition or description with mathematics as the subject. To *do* mathematics is to do research, which means to create mathematics as a poet creates poems.

When mathematicians do mathematics they publish it in journals. But what they write *is* mathematics and is not *about* mathematics. They do not write exposition. And the mathematics research they write is read, if at all, only by other mathematicians. Consequently, there exists only a sparse literature about mathematics and this is one reason that otherwise knowledgeable people are completely unaware of the existence of hundreds of mathematics research journals and of the thousands of mathematicians who publish in them.

Moreover, most of this literature is trivial because it is written not by mathematicians but rather by outsiders who fail to understand either mathematics or its nature or why people would want to do it. Occasionally, a real mathematician, for one reason or another, will produce a book *about* mathematics. But, invariably, he will confess up front that what he is doing is not his real work. Even though he knows the expository book will reach more people than all his research papers combined, the mathematician will assure the reader that the book possesses little value when compared with, say, his most recent three page article published in the *Journal of the American Mathematical Society.*

A famous mathematician named G. H. Hardy once wrote a book about mathematics. He was, however, made so contrite by the process that at the outset he apologized to the reader for having written. He, in fact, called the book *A*

Mathematician's Apology. In the first sentence he confesses his sin. He writes[1]:

> It is a melancholy experience for a professional mathematicians to find himself writing about mathematics. The function of a mathematician is to do something, to prove new theorems, to add to mathematics, and not to talk about what he or other mathematicians have done. . . . There is no scorn more profound, or on the whole more justifiable, than that of the men who make for the men who explain. Exposition, criticism, appreciation, is work for second-rate minds.

Hardy's *Apology* was published in 1940 but his attitude toward mathematical exposition continues to be pervasive. Books about mathematics are still considered by mathematicians to be work for people who are, at best, of the second rank. In fact, mathematicians consider the writing of a book about mathematics to be *prima facie* evidence of the author's failure as a mathematician. The mathematicians believe no one capable of sustaining a career *doing* mathematics would waste his time writing about mathematics. Only one who has run dry of original ideas or who, like Hardy, believes his mathematical life "is finished" qualifies to write books of mathematical exposition. And, even then, the fallen mathematician must somewhere in the writing repent of the sin of having written.

Consequently, outside of the closed circle of professional mathematicians, almost nothing is known of the true nature of mathematics or of mathematics research. And nothing whatever is known of the force that moves people to do mathematics in the first place—the attractive pull of what Hardy called the "purely aesthetic" quality of mathematics. Nor do most people know what Bertrand Russell described as the beauty of mathematics—"a supreme beauty capable of a stern perfection such as only the greatest art can show."[2]

Nothing lives further from the intellectual experience of members of the educated public than the notion that mathematics can have aesthetic value. It is remote both to those who are familiar with mathematics and to those who are not. Engineers and scientists, who use mathematics routinely in their work, see it only as a tool. Mathematics, to them, has no more charm than does a microscope or a cloud chamber. Mathematics simply helps them through a day's work. And the humanists, of course, think of mathematics not at all. Having endured years of required schooling in mathematics where the subject was presented to them as something dead as stone and dry as earth and forever separate from their own interests, the humanists have vowed never again to allow it in their presence.

Mathematicians may think of themselves as artists—which indeed they are—but the artists do not. When Diego Rivera painted a picture of a poet he produced an abstract painting which emphasized the poet's role in handling ideas and symbols. But when he painted a mathematician—a person who truly deals with ideas and symbols at the highest level—he brought forward a realistic picture of a thin man wearing rimless glasses. Rivera shows the mathematician as a man grave, awkward, and absorbed—everyone's idea of an accountant.

The invisibility of mathematics seems particularly paradoxical when matched against the size and scope of the activity in mathematics research. The current combined membership list for the three national mathematics organizations contains more than 50,000 names. Of these, more than 25,000 belong to the American Mathematical Society, which is the national organization primarily concerned with mathematics research. There are more than 1,500 mathematics journals in which appear annually at least 25,000 research papers. The *Mathematical Reviews,* the major journal publishing abstracts of mathematics papers, lists more than sixty clearly defined areas of mathematics activity. Although mathematics itself is 2,500 years old, more has been created in the last fifty years than in all the previous ages combined. Moreover, universities abound with mathematicians. On any campus the mathematics department will be the largest, or nearly the largest, academic department. Mathematicians are as numerous as physicists or economists but are far less well-known or influential. On campus, mathematicians are ubiquitous. They are also invisible.

Each mathematician lives with the invisibility paradox. Fresh journals pour into the department library bearing papers that must be read. Colloquium speakers come and go leaving in their wake new ideas that must be studied and mastered. New theories concerning subjects like fractals and catastrophes must be examined. Old conjectures like the one of Bieberbach or the problem of the four colors are settled and their proofs must be learned so that their insights can be applied to other problems and to other areas. The mathematician struggles to keep up and not to go mad. He produces new mathematics of great beauty. And if he is lucky he produces something of importance and of permanence. Yet no one outside his set of mathematical colleagues pays him or his work the slightest regard.

How can this be? What is there about mathematics that compels so many men and women to work at it with the fervor of dedicated artists and yet keeps it simultaneously outside the experience of the rest of intellectual society? What are the implications of this paradox? And what can be done about it?

This book provides partial answers to these questions: questions which are not easy to answer.

The way to proceed is to write *about* mathematics—

about what the subject is and why people do it—and to write without apology. The alternative is to write *mathematics* with all its logic and precision and carefully stated hypotheses and drawn conclusions. But then the book becomes inaccessible to the audience I most want to reach— educated people who are comfortable with art, literature, and music but who are remote from mathematics. It is this group of aesthetically sensitive people who can be most naturally touched—as the mathematicians are touched— by the aesthetic quality of mathematics. The mere fact that this needs to be done—that it has not already happened— provides *prima facie* evidence of the failure of the present system of mathematics education.

Once the secret is out and the art-appreciating public learns that the keys to mathematics are beauty and elegance and not dullness and technicality, perhaps there will be pressure for teaching reform in the direction of increasing the aesthetic component of mathematics education. Such pressure for change—if it comes at all—will come only from a newly informed and previously nonmathematical public. There is no chance, I assure you, that reform along these lines will be suggested either by the mathematicians or by their best customers, the academic engineers and scientists. And pressure for reform must come from somewhere. The failure of the present system is too obvious and too clear-cut.

If war is too important to be left to generals, then, for analogous reasons, mathematics education may be too important to be left to mathematicians.

It is not easy to write properly about mathematics. There is, first of all, the motivating force for mathematics which is *beauty,* and then the goal of mathematics which is *truth.* And finally, there is the importance of mathematics resulting from what the mathematical truths tell us about *reality.* To write about mathematics you must deal with each of these: beauty, truth, and reality. Thus, you must squarely face three of the four issues of classical philosophy. And, if you want to talk about the teaching of mathematics, about why it fails and what can be done to improve it, you are led to the fourth philosophical issue: *ethics.*

To write about mathematics you must face all four and the difficulty of doing this partially explains the dearth of quality books of mathematical exposition. Writing a paper on Banach spaces of analytic functions is far easier. Easier, that is, if you are a mathematician.

But mainly you must face the mathematics itself. How do you write about mathematics without the subject forcing on you a level of precision which will make the book unintelligible to nonmathematicians? The answer is that you describe the mathematics through metaphor and analogy exactly as mathematics itself is used by mathematicians to describe reality. When you must choose between precision and intelligibility you simply abandon precision. If you want to describe a smooth curve you merely say "smooth" and go on. You do not feel obligated, as the mathematicians invariably are obligated, to first define the notion of function and then the graph of a function and, finally, to argue that the existence of the derivative forces the graph to be smooth. You just say that the curve is smooth like an oval and not sharp-cornered like the rim of a snowflake. All those people who read Rousseau, listen to Beethoven, and appreciate Picasso will understand you just fine.

Mathematics is precise or it is nothing, and mathematicians' commitment to precision makes them wary of books of exposition. At the beginning of his fine expository book, *The Problems of Mathematics,* the mathematician Ian Stewart[3] concocts an imaginary interview between himself and a television talk-show host. In the interview,

Stewart is reluctantly convinced that, in order to make his mathematics intelligible to the man in the street, he must tone down his natural tendency to be precise. Stewart, therefore, provides an interesting variation of the traditional apology mathematicians produce whenever they consent to talk to the nonmathematical public. G. H. Hardy wrote mathematics exposition because he was over-the-hill and he believed himself unsuitable for anything more important. Stewart says simply that the devil made him do it. Or, what is the same thing, a T.V. talk-show host did.

But in the present book, neither the devil nor my own inability to do mathematics leads me to imprecision. Rather, I am imprecise because I want to be understood.

I will also frequently speak in generalities—particularly when I use the term "mathematician." By mathematician I mean, more or less, those academicians who hold tenure-track positions in research university mathematics departments. Later, I will use the term "applied mathematician," which will mainly refer to professors in those same universities—perhaps outside the mathematics department—who participate in the "applied mathematics process." The phrase will later be made clear but, for the moment, think of an applied mathematician as a person who uses reasonably high-level mathematics to solve real-world problems.

I do not believe in "collective guilt" because to do so necessarily forces a belief in the nonsensical notion of "collective innocence." So when I write a phrase like, "Mathematicians loathe imprecision," you should interpret it to mean: "Most of the tenure-track professors in research university mathematics departments greatly dislike to write or to speak imprecisely."

I assume that the acknowledgment of my own imprecision allows me to use the shorter phrase. All except the mathematicians will understand it anyway.

Robert Frost once lived in Vermont. He lived among farmers. While they worked the land he worked at poetry. The farmers broke ground with plows drawn by horses. Frost drew verse straight as furrows across blank pages. Frost and the farmers walked the same fields and saw the same dark woods. But Frost was touched in a way the farmers were not. He saw something the farmers could not see and could not appreciate. Robert Frost saw abstraction and metaphor where the farmers saw only reality. A patch of old snow was for Frost the newspaper of a forgotten day, a tuft of flowers spared by a mower became a symbol for shared values, snowfall in a dark woods turned into an experience of religious dimension. An ordinary pasture becomes an irresistible invitation[4]:

> I'm going out to clean the pasture spring;
> I'll only stop to rake the leaves away
> (And wait to watch the water clear, I may):
> I shan't be gone long.—You come too.

Robert Frost saw in ordinary things values the farmers did not see. He saw because he knew how to look and he understood, as do all true artists, that it is metaphor and symbol, and not plain reality, that is memorable and significant. Mathematicians, like poets, see value in metaphor and analogy. Similar to Frost, they draw their metaphors in lines across blank pages. But the lines they draw are made, not only of words, but also of graceful symbols: summations and integrals, infinities turning on themselves like self-swallowing snakes, and fractals like snowflakes that, as you blink your eye, turn to lunar landscapes. Mathematicians write their poetry with *mathematics.*

The purpose of this book is partially to explain what it is the mathematicians see and why it has value. And to share with the nonmathematical world an approximation of the deep sense of aesthetic pleasure that lives in the world of the mathematician.

In order to learn to see we must, like Robert Frost, go into the field. This time the field is not real but rather is the field of mathematical abstraction. But once we learn to view it properly we will see it glow. We need only practice and some new ideas.

We should go out together. It will be just a stroll, low-keyed and leisurely. I'm going out to fetch some ideas. You come too.

CHAPTER 1

The Unexpected

nce I knew a man who had been shot in the head with an automatic pistol. One day he strolled to a store to buy a bottle of Chianti and walked in on a robbery in progress. The robber turned to the man, raised his pistol, pointed it at the man's face, and fired. The robber shot him just like that—quickly, unexpectedly, and without warning. The shot was a bit off-angle and the man survived the shooting with only a permanent crease in his skull and a perpetual dislike of red wine. And a bright memory of a moment when time stood still.

"It is true," he told me. "Between the finger on the trigger and the round black hole between the eyes, there lives a different kind of time."

There are, for all of us, moments so rare as to bring with them a different kind of time. An event occurs—less dramatic perhaps than a shooting—but so charged with emotion and intensity that one's biological clock stops, the background fades away, and the thing itself is seen frozen and close up as through a zoom lens. You can hear the turn of the key as the scene locks itself deeply inside your brain.

Instantly, there is a kind of reverse *déjà vu*. You know that, for as long as you live, the moment will recur and each time you pull it from your memory it will gleam like a gold coin from a velvet case.

Of course there is more recurrence if, like the shot man, you manage to live through the moment. But the intensity and the charge and the quelled time are there whether you survive the event or not. James Dickey, in his poem *Falling,*[1] described the deathfall of an airline stewardess who was swept through an emergency exit door that suddenly popped open in the night sky over Kansas. Time's flow was staunched enough for her that as she plunged to her death she had "time to live in superhuman health." And as she planed the night air and fell past the moon she disrobed methodically piece by piece. When they found her three hours later impressed in the soft loam of a farmer's field she was stark naked.

Many of these singular moments are associated with public events. You remember exactly where you were when you learned of the assassination of John Kennedy or the death of Franklin Roosevelt. Even though you may have been only six years old, you will remember forever the bombing of Pearl Harbor. On that Sunday afternoon the grown-ups gathered in the parlor and talked quietly and soberly. Something had frightened them. Badly. And when the grown-ups were frightened, so were you.

But there are personal, non-life-threatening moments of the same kind. Once, many years ago, a pass was thrown to me in a high school football game. I can blink my eyes and bring the scene back. I'm running in the end zone, as wide open as a beach umbrella. The ball comes to me on a soft lazy spiral. No thrown football was ever more catchable. This ball comes in lace up. It smacks gently into my palms. Spectators leap to their feet and cheer. I cradle the ball and tuck it away. Then, inexplicably, the ball pops loose and I am on my knees, sliding on the grass, scrambling after the ball. There is no sound now. Only bright light and green grass and the tumbling football. I stretch desperately for the ball as if, by touching it, I can set the thing right, save the touchdown, win the game. But the ball rolls away.

Hardly a week goes by but what I still reach for that football. I didn't touch it then and I will touch it never. But I go on reaching.

What all these events had in common is their *unexpectedness*. You do not *expect* the pointed gun or the blown-open door. You did not *expect* to hear of the assassination or the bombing or to suddenly lose the treasure of the caught football as quickly as you had gained it. Each event is as completely unexpected as a ghost at the top of the stairs. Or the existence, in an ordinary classroom, of a profoundly beautiful object called pure mathematics.

THE CLASSROOM

All of us have endured a certain amount of classroom mathematics. We lasted, not because we believed mathematics worthwhile, nor because, like some collection of prevailing Darwinian creatures, we found the environment favorable. We endured because there was no other choice. Long ago someone had decided for us that mathematics was important for us to know and had concluded that, if the choice was ours, we would choose not to learn it. So we were compelled into a secondary school classroom fronted with grey chalkboards and spread with hard seats. A teacher who had himself once been compelled to this same place stood before us and day after day poured over us what he believed to be mathematics as ceaselessly as a sea pours forth foam. The room in which we sat was a dark and oppressive chamber and we thought of it then and now as Herman Melville thought of the Encantadas: only in a fallen world could such a place exist. To us, it might as well have been filled with cinders or the muck you find at the bottom of caves. For it was an atmosphere suitable only for tortoises and lizards and spiders. Like the Encantadas, the mathematics classroom was a place to be shunned by humans. On the Enchanted Isles, the chief sound of life was a hiss. In this classroom the sound was a moan.

In that room, and in the other where we had precollege mathematics thrust upon us, there were three clearly defined groups of people. The first two groups were composed of students. One of these was the collection of students who had somehow, even at the early age, showed some interest in and an aptitude for science. The other group consisted of those students who had not. Most of the students in this second group—when they were allowed to study what they wished—later pursued courses of a more humanistic nature. They became the exact opposite of scientists and these people are now generally classified under the name "humanists."

The third group in that chamber had only one member: the teacher. This person had three characteristics which we found were more or less common to the secondary school mathematics teachers that were to follow: he did not like mathematics, he did not understand mathematics, he did not believe mathematics important.

That he did not like mathematics was clear from the beginning. His lack of fancy for the subject he taught came to us, not from his negative comments about it, but from his transparent absence of passion for mathematics. Even then we knew that when you spoke of a thing you truly liked some of your enchantment with it was communicated to the listener. In fact, communication was so easy it often got us into trouble because the things we were passionate for in those days were trivialities like sports or popular music or movie stars. The grown-ups understood only too well the depth of our attraction to these petty things and they spent a great deal of their time trying to change our interests. Passion, we knew even then, is too easily communicated. But from our teacher we heard nothing of mathematics except basic facts. When he spoke to us of mathematics he spoke with neither ardor nor metaphor. He taught mathematics on weekdays with less enthusiasm than he showed on Saturday when he mowed his lawn. We *knew* he did not like mathematics. But we did not hold that against him. We did not like it either.

The knowledge that he did not understand the subject came to us more slowly and became certain only when we had studied more advanced mathematics. We could then —as earlier misunderstood notions became clear—look back on what he had told us about mathematics and pinpoint exactly the shallowness of his understanding. But we saw unmistakable signs of his ignorance even as he taught us. Mostly they showed through his fumbling and fearful responses to elementary questions that he could not answer: Why is it *really* that the product of two negative numbers is positive? How do you know those two bisectors intersect *inside* the triangle? What is the *next* number after ½?

His reaction to each exposure of his ignorance was always the same. First came the fumbling stage where he tried to talk the question away by saying whatever popped into his head the way a politician tries to explain away a vote against social security to a group of elderly constituents.

When that failed, as it invariably did, and he saw that it had failed, he became fearful. And when we saw the fright flash in his eyes we knew that this subject—mathematics—terrified our teacher deep down exactly as it terrified us. But there would be only a flicker of fear and then he would become authoritative and, in one way or another, denounce the attitude of the student who asked the question for having the temerity to ask it.

We were not, however, surprised at his ignorance of mathematics. Neither are the students who now sit in the same place and see the same ignorance in their teachers. At that stage in your life you expect no one to understand mathematics. It is just something you endure for as long as you must.

Of course, you also do not expect the person who teaches you mathematics in secondary school to think it is valuable. Why should he? No one else you know does. Your parents live their lives without mathematics and so do your parents' friends. At no place you go—the grocery store, the dentist, the movies—do you see the slightest bit of evidence that any mathematics beyond the basic ability to do simple arithmetic is anywhere needed in "real life." Mathematics is mentioned neither in the newspapers nor on television. At no time has mathematics ever been—within your range of hearing—a subject of conversation.

Naturally, your teacher *tells* you almost every day that mathematics has value. But you know that he does not believe it. And he knows that you know. It is just another shared fiction. Like the notion that the poor shall inherit the earth.

We went on, this unhappy band of three. We took a succession of secondary school courses. And, at each stage, the separation between the future scientists and the future humanists became more and more sharply defined. The "scientists" were told they had to continue with mathematics to do what they were ordained to do later in college. So they grimly hung on to the mathematics curriculum, determined that while the subject might never be understood, it could at least be *learned*. As the curriculum advanced, the "humanists" became more and more ignored by the mathematics teachers and advisors, and were allowed—even encouraged—to drop out of the mathematics sequence.

The mathematics teachers, as might be expected, continued. They came before us one by one, uninspired and uninspiring, as identical as dominos.

ACCIDENTS

Most of the mathematicians of my generation will admit they came to their profession by accident. Few of us set out to be mathematicians. Indeed, because all of our secondary school and beginning college courses in mathematics had been taught from the point of view of the utility of the subject, none of us knew that such a profession was possible. Because we had shown some competence in high school mathematics we had been advised to become engineers or scientists. Our teachers, having themselves been taught mathematics only as a tool for applications, had no notion that someone good at mathematics might consider becoming a mathematician. Indeed, then—as now—there was almost no notion among secondary school teachers of mathematics that a profession called "pure mathematics" even existed.

So, we began, in college, to study engineering. We took calculus, differential equations, and linear analysis—all of these being courses emphasizing specific mathematical techniques used in engineering applications. Then, just by chance, we took a course outside our curriculum. We enrolled in a postcalculus course in analysis. And, unexpectedly, we encountered pure mathematics. We were struck by the subject, like Saul on the road to Damascus.

Nothing in our backgrounds had prepared us for the aesthetics of mathematics. We saw, for the first time, a professor who treated mathematics with reverence, who wrote symbols on the blackboard with great care as if they mattered as much as the information they contained. We heard a mathematical result described as *elegant*. And we saw that it was.

It was a moment of great discovery. It was as though we had lived all our lives in the hold of some great ship and now were brought on deck into the fresh air. And we saw the unexpected sea. We felt, with W. B. Yeats[2]:

> All changed, changed utterly:
> A terrible beauty is born.

The moment was worth the wait but we would be forever aware that we had come to it entirely by chance. And, along the way, there had been many dropouts.

TRIVIALITIES

In secondary schools, we had taken a succession of courses which consisted mainly of the repetitive manipulation of mathematical symbols. Each dull course had always been justified on the grounds that, once it was mastered, we would have at hand some kind of a machine for transportation. This machine, they told us, would transport us from the classroom to some vague place called the real world.

And the word they used most frequently to indicate the link between this place and mathematics was "applications."

As an application of algebra, we had computed the age of a farmer who is twice as old as his son will be in six years if the farmer is now three times as old as the son. Using geometry, we had measured the size of plots of land along the Nile River as did the ancient Egyptians. Trigonometry had been "applied" to the determination of the heights of an unending succession of trees, once certain angles were measured.

But we didn't care much about the age of fictitious farmers or the size of Nile River bottom land or the heights of trees. Even those of us with scientific interests were not persuaded there existed any connection between such calculations and the difficult notions we believed lay ahead in physics or chemistry. We were naively certain that, in order to get at the central core of scientific truth, one needed a tool at least as complicated as a jackknife. And all we had seen of mathematics was its application to a dull sequence of trivialities.

For the others, the nonscientists, mathematics was even more a test of endurance and will. What interested them about farms and rivers and trees had nothing to do with calculation. They cared for pasture springs, the sound of wind through the pines, the slant of sunlight on bright water. Mathematics was to them as far removed from aesthetics as is a dusty tool box from a polished violin.

These humanists waited. They patiently suffered through the minimal amount of mathematics instruction and took silent vows that, when it was over, they would never again allow the subject to be brought into their presence.

The rest of us continued haltingly through the curriculum. We sidled up to each mathematics course the way a kid with skinned knees limps toward his two-day-old bicycle. Mathematics might well be a vehicle as our teachers said, but it had, so far, taken us nowhere we wanted to go. So far, mathematics had only thrown us on our knees.

College calculus came next and with it the defeat of all expectation. Through high school, calculus had been dangled before us like some golden ring. Our teachers had talked of it as if it were the capstone of all mathematical knowledge. There was never the slightest indication from them that analysis beyond calculus existed. "When you get to calculus," they told us, "you will see what mathematics is really good for."

What we saw was yet another course in manipulation taught from the point of view of physical applications. Derivatives were presented as velocities and integrals as areas. As far as we knew, the difference between a Riemann integral and an antiderivative was that one of them was evaluated between limits. And for the purposes of the course no further knowledge was needed.

We waited a full year before we, unexpectedly, discovered the existence of pure mathematics and learned that elementary calculus had been given to us exactly the wrong way around. The truth is that there is a mathematical notion called the derivative of a function which represents a rate of change of one variable with respect to another. Sometimes the derivative can be interpreted as a velocity. But derivatives also have other interpretations as, for example, interest rates or probability densities or population growths. So what one should study are the derivatives themselves and have the applications appear as special cases.

One full year passed after elementary calculus before we learned the true relation between Riemann integrals and antiderivatives. We discovered that the connection between these very different notions lies at the very heart of the subject and that it is one of the genuinely great creations of the human intellect. We saw that the connecting argument is a thing of great beauty. Suddenly we understood that mathematics has an aesthetic value as clearly defined as that of music or poetry.

CHAPTER 2

Pure Mathematics

athematics falls roughly into two major divisions: pure mathematics and applied mathematics. The separation between these two subdivisions is inexact and mathematicians argue continually over the placement of the boundary that marks the end of one and the beginning of the other. The arguments are futile and one can imagine a sequence of pictures showing the transformation of the most applied of applied mathematicians on the left into the most pure of pure mathematicians on the right. At no place in between could you say, "Here is where the person becomes a pure mathematician," just as in the familiar sequence of evolutionary pictures one cannot point to the exact moment at which the ape becomes human. But the difference between the ape and the human is clear and so is the difference between the extremes of applied and of pure mathematics. In the university, there is no hostility more bitter than that existing between the mathematician who is clearly on one side and the mathematician unequivocally on the other.

Once, years ago, I was asked by my university's chief academic administrator to chair a committee whose charge was to bring together, in some common departmental setting, our separate programs in pure and in applied mathematics. I began my work by visiting individually the major campus figures in the mathematical sciences. My first visit was to the university's most distinguished pure mathematician. I explained the committee's charge and listed for him what I believed to be sound academic and administrative reasons for merging the applied and the pure mathematics programs. But what I said had as much effect on him as a fundamentalist baptist sermon has on an archbishop. He told me: "Forget it. We will never work with *those* people. They haven't the slightest conception of what mathematics is all about."

Two hours later, in my second interview, the most distinguished applied mathematician on campus told me the exact same thing.

I thanked each mathematician for his time and I pointed out that, since neither of them was immortal, what each would "never" do was, in the long run, of little interest to the university. "In the long run," I said, following the economist John Maynard Keynes, "we are all dead." And I told them that the ultimate merging of the two disciplines was inevitable since it was clearly in the best interest of the university, of the students, and of mathematics itself.

But I was younger then, and naive. And, although each of the two mathematicians with whom I spoke retired long ago, the disciplines today remain separate and the hostility continues.

Basically, pure mathematics is mathematics for mathematics' sake and applied mathematics is mathematics for something else. And the "something else" of applied mathematics is invariably some aspect of reality.

There exist, for each mathematician, two clearly defined worlds. The first is the world of mathematics and the other is the world of reality. The real world is just what you think it is. It is the world of sense experience—the world you see and feel; the world in which you live. What lives in this world are people and places, sunsets and pasture springs, atoms and oceans. Death lives here also, as does disease and disaster. This is the world of nature—good and bad—and it is the desire, and the necessity, to understand and to control this world that has caused humankind to invent and to theorize. Science sprang from this necessity. So did magic and religion. And so did mathematics.

The mathematical world, on the other hand, is a world of ideas. The things that live in this world are mathematical objects like numbers and analytic functions, matrices and differential manifolds, sequences and topological spaces.

There are, to be sure, deep philosophical issues which lie beyond the scope of this book. It is not my intention to attempt to describe the exact nature of physical reality. Nor can I avoid all apparent contradictions. The basic notion is simply this: the mathematical world lives in the mind; the real world lives outside it.

If you think of the mathematical world as being represented by a large rectangle, one way to divide it would be into "countries," each having the name of one of the sixty areas of mathematics identified in the index to the *Mathematical Reviews*. This classification gives an abstract world broken into pieces with names like "category theory," "complex variables," and "algebraic topology" (see Figure 1).

Dividing the mathematical world into regions as I indicate in Figure 1 causes inaccuracies just as does separating mathematics into "pure" and "applied." Segments of the boundaries between the "countries" may be vague. More-

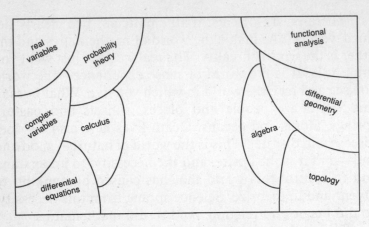

FIGURE 1. The mathematical world.

over, the areas often overlap and a particular piece of mathematics may belong to more than one of them. For example, the motion of a falling raindrop may be represented by what is called an ordinary differential equation which belongs, naturally, to that portion of the mathematical world called "differential equations." However, the elementary nature of this equation causes it also to fall within the region of Figure 1 labelled "calculus," showing that these two parts of the mathematical world have a nonempty intersection. Moreover, the "differential equation" region will also intersect the areas called "real variables" and "complex variables" as well as several other areas shown (and not shown) in Figure 1. But even though the diagram is inexact, it gives a convenient metaphor for the mathematical world and we will return to Figure 1 later in the chapter called "Applied Mathematics." (Incidentally, there is no subfield of mathematics listed in the *Mathematical Reviews* called simply "calculus." I have included this area in Figure 1 because of its familiarity. Also we should notice that there are no raindrops in the mathematical world. Raindrops live in the real world. It is the model of the raindrop—in this case a differential equation—which belongs to the mathematical world. Mathematicians study models. They do not study raindrops.)

A mathematician, like everyone else, lives in the real world. But the objects with which he works do not. They live in that other place—the mathematical world. Something else lives here also. It is called *truth*.

TRUTH

When Jesus Christ stood bound before Pilate, he said: "Everyone that is of the truth heareth my voice." And Pilate replied: "What is truth?"[1] Then Pilate washed his hands.

Mathematicians are not philosophers. They do not spend their professional time wrestling with the classical philosophical questions. Neither do they, like Pilate, shrug them away. Mathematicians know beauty when they see it for that is what motivates them to do mathematics in the first place. And they know where to find truth.

Mathematicians *do* mathematics. And when they do mathematics they deal with objects they have created. These objects are abstractions and have no existence outside of the imagination of the mathematician. They are endowed by their creator, the mathematician, with certain properties. From these assigned properties, using the laws of logic and the rules of mathematics, the mathematician deduces other properties. The objects of pure mathematics are *completely perceivable* for they possess only the properties they have been assigned and the properties that can be deduced. If the work of deduction is done properly and without mistakes, the derived properties will be as close to absolute truth as humankind is likely to get.

A mathematician named Alfred Renyi said: "Is it not *mysterious* that one can know more about things which do not exist than about things which do exist?"[2] Alfred Renyi was a Hungarian mathematician who died in 1970 at the early age of 49. He left behind a considerable body of research in complex analysis and probability theory. Renyi's work is significant and, I believe, not nearly well enough known. It is work for which I have great respect. Renyi also wrote *about* mathematics and he wrote about it very well. He was a rare bird in the aviary of mathematicians because he wrote mathematical exposition without apology and with the same seriousness with which he wrote mathematics. For this, I respect him even more. I do not, however, always agree with him.

The italics in the above quote are mine and I placed them there to call attention to the word "mysterious," because I think Renyi has misused it. A more accurate statement would be: "It is *commonplace* that one can know more about things which do not exist than about things which do exist."

Consider, for example, the question of the guilt or innocence of the English King Richard III. Did he, or did he not, order the murder of the two princes in the Tower of London?

There are, of course, two kings called Richard III. One of them was a real person who was born in 1452, died in 1485, and reigned from 1483 to 1485. The other Richard III is an abstraction created in the mind of William Shakespeare and described by him in his play *The Tragedy of King Richard III.*

The historical Richard III, like most real things, is vague and what we know of him is inexact. Four centuries after his death we are still uncertain whether or not he was a monster who put small children to death or if he was an innocent who has been given a "bad press" by rumor, innuendo, and inexact knowledge. Most people—following Shakespeare's lead—believe him a monster. Josephine Tey, on the other hand, believes him innocent. And in her novel *The Daughter of Time,* she has her protagonist say of him[3]:

> From the police point of view there is no case against Richard at all. It isn't that the case isn't good. Good enough to bring into court, I mean. There, quite literally, isn't any case against him at all.

On the other hand, there is no doubt concerning the imagined Richard. In Act IV of the play Richard sends Tyrrel to kill "those bastards in the Tower." When Tyrrel says he will murder them, King Richard purrs villainously: "Thou singest sweet music."[4]

But, while no mystery remains about the guilt of Shakespeare's Richard III, we have no "truth" about the other Richard. His guilt or his innocence appears as uncertain as was the weather the day your grandfather was born.

What *is* mysterious is that we can ascertain *anything* we are willing to call "truth" about real things. For any attempt at determining this kind of truth, sooner or later, requires the use of our senses. This is the case whether we are simply looking outdoors to determine today's weather or we are using the most sophisticated instruments of modern experimental physics to comprehend the erratic behavior of the particles which make up the atom. Ultimately, we must *look* at the thing we want to understand—either directly with our eyes or indirectly by way of the output of cloud chambers or electron microscopes. None of us doubts that our senses have, on more than one occasion, proved to be fallible. A *single* failure is enough to make us question whether any "fact" determined by them deserves to be called truth.

At the level of ordinary experience, the fallibility of human observation provides material for motion picture plots, party games, and freshman psychology experiments. In these settings, several people simultaneously observe the same event. Later, they are asked to describe what they have seen. The inevitable discrepancies in their reporting of the observed data provide material for the plot or the game or the experiment. (Social scientists call this phenomenon the "Rashomon effect" after the famous 1950 Japanese film.)

These inevitable—and expected—discrepancies in the testimony of eyewitnesses also keep lawyers and courtrooms busy and, in our litigious society, have led to the establishment of the pseudo-scientific profession of "expert witness." An expert witness is a person of reputation and authority who gives to the court and the jury an interpretation of the data which are pertinent to the trial and which have already been interpreted by ordinary witnesses. The expert's testimony is, presumably, rational and disinterested. The trouble is that each side has one. My expert says one thing. Yours says exactly the opposite. And the "truth" becomes whatever the jury decides.

On a deeper level—the level of science and scientists —truth remains elusive. In science, the approach is essentially a pragmatic one. Does the "truth" agree with observation and experience? If not, then there is a "better truth." The scientists' job is to find it.

Aristotle told us that uniform motion was possible only when there existed a force to maintain it. This "truth" essentially agreed with observation until Galileo made more precise observations of falling bodies and told us the exact opposite. But Aristotle's scientific "truth" had centuries of validity. Ptolemy, in about 100 A.D., explained planetary motion starting from the hypothesis that the earth was fixed in space. He concluded that the other planets move around the earth on a combination of circular orbits called epicycles. This "truth" stood for more than 1,500 years until Copernicus and then Kepler showed that the planets, including the earth, move not in circles or combinations of circles but in ellipses. Kepler's work led to Isaac Newton and the *Principia,* said to be[5]

> . . . the culminating intellectual creation of the seventeenth century, the clockwork universe of Newton in which marbles and planets rolled about as a result of the orderly interplay of gravitational forces, in which motion was as "natural" as rest, and in which God, once having wound the clock, had no further duties.

This "orderly interplay" resulted from Newton's famous inverse square law of gravitational attraction and stood unchallenged as *the* explanation of planetary motion until the twentieth century when Albert Einstein showed that what is fundamental is not force, but rather the geometry of space–time. And so it goes. One scientific truth was superseded by another, and each new truth came from what Jacob Bronowski has called "the habit of testing and correcting the concept by its consequences in experience. . . ." And, according to Bronowski, ". . . the habit of simple truth to experience has been the mover of civilization."[6]

Perhaps. It has certainly been the mover of science. But the issue is whether or not this habit will ever lead us to something that finally can be called truth. Or is it that science must unendingly peel away layer after layer of multireality and with each successive layer it is necessary to accept a new and different truth?

There is more to the world than *science* can ever know. In *The Golden Bough,* Sir James George Frazer said[7]:

Yet the history of thought should warn us against concluding
that because the scientific theory of the world is the best that
has yet been formulated, it is necessarily complete and final.
. . . as science has supplanted its predecessors, so it may hereaf-
ter be superceded by some more perfect hypothesis, perhaps by
some totally different way of looking at phenomena—of regis-
tering the shadows on the screen—of which we in this generation have no idea. . . . The dreams of magic may one day be
the waking realities of science.

Indeed. And one thing is sure. If you want truth that
will not change with tomorrow's observations you do not
look for it in science. You look for it in mathematics.

Mathematicians know the answer to Pilate's question:
"Truth, Mr. Roman Governor, is what you find at the end
of a correct chain of mathematical argument."

RESEARCH

Technically, to do research in pure mathematics
means to produce new mathematics. Realistically, it means
to produce mathematics that is new and that is simultaneously significant. For the only research that *counts*—at
least for academic mathematicians—is that research which
is publishable in what are called "reputable and refereed
journals." Only original research that is somehow significant is deemed by these journals to be worthy of
publication.

So, in a technical sense, you are doing research in mathematics whenever you are producing mathematics that is
new to you. Whenever you solve a problem in pure mathematics—assuming that you do not know the solution be-
forehand—you are, in a minuscule sense, doing research. If
you have invented the problem yourself and then solved it

yourself, your work is close to what mathematicians actually do. If the problem you have invented and solved is a *new* problem, then you are doing actual research in pure mathematics. If the new problem is also *significant* then what you are doing is publishable research.

It is easy to find a problem in mathematics that is new —a problem whose solution does not exist in the vast collection of books and journals that constitute the mathematics literature. For example, consider the following: Let n denote the number of words printed in last Sunday's *New York Times*. Multiply n by the square root of 17. Call the new number m. Find the one millionth digit in the decimal expansion of m. This problem is undoubtedly new and if you solve it you will be producing a mathematical result that is not yet known. But you will also be producing something of little interest and of even less significance. It will not be publishable mathematics.

Any professional mathematician can, from his experience and his knowledge of the mathematics literature, produce a problem whose solution will be not only new but significant. This is also easy. What remains hard is to find a problem which possesses both the features of originality and significance and simultaneously falls within the scope of the researcher's power to find a solution. Most of the people who publish in mathematics journals are those who have learned to do this routinely and well. The *good* mathematicians are those to whom the *problem* is paramount. And, once attracted to a problem of significance and elegance, they learn or create whatever mathematical methods are necessary to solve it. *Mediocre* mathematicians, on the other hand, are characterized by their tendency to use only the mathematics they already know and to search for problems that can be solved by these *methods*. This means, naturally, that much of their work seems forced and the results seem technical and uninteresting. Since there are far more mediocre mathematicians than there are good mathematicians, most of the 25,000 research articles published each year are uninteresting and have little permanent value.

Thus, mathematics has at least one characteristic in common with every field of endeavor, namely, that the mediocre people vastly outnumber the good people. This is not a pleasant characteristic nor is it surprising. It is, in fact, almost the definition of the word "mediocre."

Ordinarily, a mathematician does his work alone. Research articles written by two mathematicians are not uncommon but they are greatly in the minority. And articles having more than two authors are quite rare. This should be contrasted with published research in the physical sciences where multiple-authored papers are the norm. In chemistry, for example, you commonly see research articles bearing the names of five or more authors. Here, research is clearly a team effort and the "major investigator" often serves as a kind of "idea person/administrator" whose primary responsibility is to get the project going and see that it is properly funded. The actual research—usually experimental in nature—is carried out by the "team" which may contain a number of graduate students as well as senior-level personnel.

A mathematician, however, almost always works alone. Unlike the scientist who needs sophisticated laboratory equipment to do his research, the mathematician needs only paper and pencil. When chemists work you find them in their laboratories busily reading dials, observing phenomena, and directing one another in the conduct of some great experiment. When a mathematician works at mathematics he sits alone in his study staring at equations scribbled on his blackboard or at a dog-eared reprint of the research paper whose results he is trying to extend. It is quiet work, like writing poetry, and includes lots of "dead time" when the mathematician, like the poet, does nothing but sit and stare at the blank page. When you walk in on a research mathematician and find him reclining with his feet up, gazing wistfully out the window, what you say is: "Sorry, I didn't know you were working." Because he probably is.

Every mathematician nowadays has a specialty. Some of the great mathematicians of the past such as Leonhard Euler in the eighteenth century and Carl Friedrich Gauss in the nineteenth were essentially able to keep up with developments in all of mathematics. Some modern mathematicians, like David Hilbert, who died in 1943, have contributed to several areas of mathematics, and have produced new mathematics belonging to several of the "countries" shown in Figure 1. Hilbert, for example, produced important research in algebra, number theory, functional analysis, and the foundations of mathematics. (Incidentally, you should not confuse the term "algebra" used here with the subject of the same name you studied in high school. Here, the term refers to a branch of mathematics that deals with sets having names like "groups," "rings," or "fields" and possessing structures which allow the set elements to be combined in various ways. What you studied in your high school "algebra" more than likely contained only rules for the elementary manipulation of numbers.) The mathematician Ian Stewart suggests that Henri Poincaré, who died in 1912, was "the *last* person to understand fully all of mathematics."[8]

The typical mathematician now considers himself "expert" only in some particular field of mathematics. And, because of the enormous amount of mathematics that is being produced, he may restrict his research to only a part of one of the areas shown in Figure 1. Thus, he may find that, although algebra is his field of interest and the area in which he concentrated as a doctoral student, the general area is far too vast for him to master. Consequently, he confines himself to working in some particular subset of algebra, say, the theory of groups. More than likely, the subarea itself becomes more than he can handle so he specializes even more in groups possessing particular properties. As he ages and becomes better at the mechanics of the research game, while simultaneously losing his enthusiasm and ability to do creative mathematics, the specialization becomes even sharper and he concentrates more and more on less and less until ultimately he deals only with topics like "the weak equidistribution of sequences in finite commutative groups admitting nonstationary operators." What he then produces are specialized research papers which are of interest—if to anyone—only to a handful of people scattered around the world.

It follows that our typical mathematician's research papers are read by only a few people. In fact, an ordinary mathematics research paper is read by almost no one except the author and the journal "referee" who reviewed it prior to publication.

Even though no one reads a typical mathematics research paper, faculties and administrators of the major universities universally believe that everyone—the professor, the students, and the institution—benefit from the writing of the paper. In fact, no nontenured faculty member who does not regularly publish such papers will be continued at a major university.

Our typical mathematician produces mathematics "by the numbers" (no pun intended). I mean by this that he will have mastered the tools and the literature necessary for him to pursue research in his tiny subarea of the theory of groups. He is able, by virtue of his experience and training, to examine a small number of published papers and thereby keep track of what is happening in his corner of mathematics. And he can recognize the gaps in the publications and identify the places where he can extend the results or fill in the missing parts with research of his own. He can do this not because of any insight or real depth of understanding but because he has patiently and carefully mastered the "tools of the trade." The resulting research is neat, usually correct, and invariably pedestrian. He produces mathematics the way a carpenter makes a kitchen cabinet. Adequately.

Our mathematician, being typical, will be mediocre. In fact, the mathematician Alfred Adler says in his provocative piece in the *New Yorker*[9]:

> There are no acceptably good mathematicians. Each generation has a few great mathematicians and mathematics would not even notice the absence of the others. They are useful as teachers, and their research harms no one, but it is of no importance at all.

Of course, Adler's "great mathematicians" do not do mathematics the way our typical mathematician does mathematics. They do not, in the first place, confine their attention to tiny subsets of mathematics but remain aware of, and actually work in, larger areas of the mathematical world. And the great mathematicians do not work at mathematics by the numbers. The typical mathematician does mathematics the way a mediocre chess player plays his game. The chess player looks ahead and examines carefully the potential placement of the pieces on the board a move or two ahead. He says to himself: "If I move my pawn here then he will move his rook there. Or if I move my bishop he may jump his knight and then . . ." And so it goes. The analysis breaks down quickly because the number of combinations of possible moves becomes staggeringly large and the mediocre player cannot keep track of them. The mediocre mathematician arranges his "mathematical moves" in exactly the same manner. In this case, the opposition is not another player across the board but rather the mathematical literature itself. What our mathematician does is move his mathematical objects around, according to the well-established rules, in order to land on blank spots in the literature. A great chess player, on the other hand, sees not just individual chess pieces and the constraints on each piece that, like strings on a puppet, control the movements they can or cannot make. But, rather, the great player sees the "game." He somehow, by means of a mental process he neither understands nor has learned, comprehends the flow of the game as a continuum and not just as a set of discrete jumps. Similarly, through intuition and insight, the great mathematician "sees" mathematics, if not as a whole, at least in great chunks.

The great mathematicians *feel* mathematics in a way the rest of us do not. And their genius for mathematics is immediately recognizable. When Gauss was eight years old, he and his classmates were asked by their teacher to find the sum of the integers from 1 to 100. The children began laboriously to calculate on their slates. All of them (except Gauss) began $1 + 2 = 3, 3 + 3 = 6, 6 + 4 = 10. . . .$ Gauss noticed that the integers $1, 2, 3, . . ., 99, 100$ can be placed in pairs as follows: $(1, 100), (2, 99), (3, 98), . . ., (50, 51)$. There are exactly 50 such pairs and the sum of the integers in each pair is 101. Hence, the desired sum is the same as 50 times 101, which is 5050. Gauss wrote this number on his slate and handed it to the teacher. The whole process took him only seconds.

Gauss went on to become a towering figure in mathematics and anyone can make a strong case that he was the greatest mathematician who ever lived. While his insightful classroom calculation was not significant mathematics, it was an early indication of his genius. Gauss, like most of the other mathematicians in, or nearly in, his class, did his best work when he was relatively young. The British mathematician G. H. Hardy said: "No mathematician should ever allow himself to forget that mathematics, more than any other art or science, is a young man's game."[10]

And, in mathematics, there is not much you can do to attain greatness. Those men and women who attain it manage to do so through their uncanny "feeling" for the subject. There is nothing you can do to learn this "feeling." You can, to be sure, work hard enough at mathematics to make yourself into a competent mathematician, perhaps even a competent research mathematician. But you cannot learn to do mathematics the way the great mathematicians do, any more than you can learn to throw a fastball or to run 100 yards in ten seconds. You can or you cannot. If you *can,* you will know it early. If you are an adult and you are not already a great mathematician then you are not going to be.

CREATION

Earlier I wrote that to do research in pure mathematics means to *produce* new mathematics. The word "produce" as used here seems slightly awkward and it would be more natural to replace it with "create" or with "discover." But I have used produce and its variants until now to describe the work of mathematicians because it is, in fact, a continuing controversy in mathematical circles as to whether new mathematics is *created* or whether it is *discovered.* We need to examine, albeit briefly, this distinction.

The viewpoint that mathematics exists, as the physical

world seems to exist, independent of human thought and activity is a notion at least as old as the philosophy of Plato, who, in the words of Bertrand Russell, "regarded the contemplation of mathematical truths as worthy of the Diety."[11] The Platonistic view, as related to contemporary mathematics, holds that mathematics and mathematical truths already exist—out there somewhere like the far away stars—and what mathematicians do is *discover* them. These truths, say the *absolutists,* are found the way new galaxies or new chemical elements are found. The fact that they are found by thought and not by experiment, they say, does not mean that they were not there already. The absolutists simply extend to mathematics the ancient philosophical principle of plenitude: if something is conceivable, then it must somewhere exist. (We must be careful here. Mathematical objects are objects of the imagination. To say that a mathematical object *exists* means that it has been *conceived* in such a way as to be consistent with the rules and logic of mathematics. This logic, for example, does not allow the existence of a triangle whose interior angles sum to 200 degrees. Nor does there exist a natural number whose square is negative.)

The absolutist notion implies, for example, that if I invent a *new* mathematical object which consists of solutions of a certain class of differential equations—call it a rabble—and define certain operations acting on its members, then all the properties of this object suddenly pop into existence like a hydrogen atom in a steady-state universe. Instantaneously, the absolutists say, every single truth about rabbles exists. This is so, they say, even though I have *produced* the thing completely alone in my study in the middle of some dark night. Even though no other human has seen or heard of a mathematical rabble all truths

about it exist. What the mathematician does is discover them.

The second point of view claims that mathematical
structures are *created* and they have no existence independent of the person who created them or of the people who
study them. This notion meshes well with the nature of
modern pure mathematics where abstraction piles upon
abstraction and the ultimate objects of the mathematician's research are many layers removed from any possible
physical interpretation. Numbers, for example, are replaced by more general objects having analogous properties. These objects are themselves combined into "spaces"
on which certain "functions" are defined. And these functions, which are, by now, at the fourth level of abstraction
from anything which can remotely be considered as "natural," become the fundamental objects of the mathematician's interest. The Platonistic point of view must be
stretched very thin to provide independent existence to
things so estranged from the real world.

The creationist viewpoint fails to account, however,
for the recurring applicability of some of the purest and
most abstract of mathematical structures. Riemannian geometry, for example, follows with mathematical certitude
from a set of axioms exactly as does the familiar geometry
of Euclid. The Euclidean axioms are nevertheless "natural"
in that they express mathematically "facts" about lines and
points which seem obvious from our view of the three-
dimensional world in which we live. Bernhard Riemann, in
about 1850, created a set of axioms which differ significantly from those of Euclid. In particular, Riemann hy-
pothesized, contrary to Euclid, that, given a line and a
point not on the line, there exists *no* line through the point
parallel to the given line. He also adopted the axiom that

two points may determine more than one line—not a single line as was considered obvious by Euclid and his followers for two centuries. The results Riemann proved from his unnatural axioms formed a set of truths as certain as the results of Euclidean geometry. But it also seemed certain that they could have no significance beyond their value to pure mathematics. Other non-Euclidean geometries which carry the names of Lobachevski (1793–1856) and Bolyai (1802–1860) have exactly the same status: they follow logically from a set of axioms different from the axioms of Euclid and they give results which differ wildly from our observations of reality.

Following Einstein, modern physicists have discovered that actual space is not Euclidean but is rather more nearly Riemannian. This discovery caused consternation to those who believed the value of mathematics lay largely in the ability of mathematicians to extrapolate "true axioms" from the real world and, consequently, to deduce unique theorems describing reality. Morris Kline, author of *Mathematics in Western Culture,* in fact, has written[12]:

> The creation of non-Euclidean geometry cut a devastating swath through the realm of truth. . . . In depriving mathematics of its status as a collection of truths, the creation of non-Euclidean geometries robbed man of his most respected truths and perhaps even of the hope of ever attaining certainty about anything.

Riemann and the others, of course, did no such thing. What they did was *create* new mathematics in the exact way that all mathematics is created: by formulating axioms and proving theorems. The fact that the axioms are "unnatural" is of no consequence to mathematics. The additional fact that these non-Euclidean geometries provided the model needed by Albert Einstein in his 1915 theory of general relativity and that they, and not the natural geometry of Euclid, appear to be the real geometry of the physical universe is of importance to mathematicians only in that it provides an overused justification for research in the ultra-abstract world of pure mathematics. This justification asserts: *any* piece of mathematics research has the *potential* value of being applicable to the real world.

Absolutists may be disturbed that "sensible" geometries follow from nonevident axioms so that mathematics becomes more clearly an invention of the imagination and not an *a priori* set of universal truths. But creationists are unperturbed and they go on creating. The real mystery is not that strange pieces of pure mathematics turn out to be applicable to real-world phenomena but that *any* result in pure mathematics has this property.

An interesting aside to this creation/discovery controversy over the nature of pure mathematics lies in the remarkable book *The Tower and the Bridge,* by David Billington. Billington, Princeton professor of civil engineering, discusses the "fallacy" that engineering and technology are the same as applied science. He says[13]:

> Engineering or technology is the making of things that did not previously exist, whereas science is the discovering of things that have long existed. Technological results are forms that exist only because people want to make them, whereas scientific results are formulations of what exists independently of human intentions.

Billington is saying that engineers *create* while scientists *discover.* This puts the engineers, who are disdained by the mathematicians, squarely in their camp and places the scientists, to whom all pure mathematicians are indifferent but to whom homage is paid for funding reasons, with the absolutists.

Billington may be wrong but he is convincing, and when he speaks of "the centrality of aesthetics to the structural artist," and when you realize that by "structural artist" he means a certain type of engineer, the association of engineers with mathematicians becomes more firm. It would be ironic if, at the highest levels, the engineers were motivated as mathematicians are motivated. For the mathematicians think of engineers as being unworthy of serious instruction in the mathematical arts. The humanists think of engineers in analogous ways even though any arbitrarily chosen engineer knows far more about art and humanities than a room full of humanists knows about technology.

But we need to move on with our subject and we should next look at the objects that are most commonly associated with mathematicians. These objects are called numbers.

CHAPTER 3

Numbers

Mathematical concepts came originally from natural objects because mathematics developed as an attempt to understand nature. We see this most clearly in the lines and triangles and other figures of Euclidean geometry whose properties agree almost exactly with our commonsense notion of their real-world counterparts. Nevertheless, the geometric figures of Euclidean geometry are abstract objects and live in the mathematical world and not in the real world. A triangle is an object which exists in the mind of a mathematician. It has properties imposed on it by the axioms of Euclidean geometry and other properties which can be deduced from these by the laws of logic and the rules of mathematics. The fact that the properties we deduce for the mathematical objects seem to tell us something about real-world objects that look like "triangles" is a wonderful thing. But it is not mathematics. Mathematics is done in the mathematical world of ideas and not in the real world of objects.

Although they are the most fundamental of mathematical objects, the *natural numbers* are not found in nature. These numbers are the familiar counting numbers 1, 2, 3, . . . (the three dots are read "and so forth"). They are abstractions from early attempts of humans to make some sense of what was common between groups of objects that were the same "size." The number "3," for example, is what all triples have in common whether they are triples of sheep or triples of solutions to cubic equations. But this "commonness" is an abstraction and there are no "3s" in the real world. And there will be none even if you write a "3" on a sheet of paper and hang the paper on a tree. You will have failed to create a physical "3" exactly as you fail to create a dragon by painting a picture of one on the side of a barn. Dragons and "3s" are concepts. They are not physical objects. *Pictures* of dragons and *symbols* for "3s" are.

To be sure, there is something terrifyingly natural about the counting numbers 1, 2, 3, . . . which is precisely why they are called natural numbers. They were the first numbers to be used in any systematic way as mathematics began to be developed all those centuries ago. And they continue to be the first numbers each of us encounters when, as small children, we first learn the process of counting. Even after mathematics had become enormously complicated in the nineteenth century and then contained many curious objects which could also be used for "counting," the mathematician Leopold Kronecker reaffirmed the fundamental character of these numbers by saying: "God made the natural numbers; all else is the work of man."[1]

But Kronecker was by no means the first person to attach religious significance to the natural numbers. Long before, about 540 B.C., a Greek named Pythagoras founded a school partly mathematical and partly religious. The Pythagoreans literally worshipped the natural numbers and they believed the entire universe was made up of these numbers and ratios of these numbers. They assigned them names like "masculine" and "feminine" and "amicable." Once they discovered the relationship between the ratios of natural numbers and the musical scale they extrapolated this into a belief that the moving planets demonstrated the divineness of these numbers as they played "the music of the spheres." Kronecker believed God made the natural numbers. But to the earlier Pythagoreans, these numbers *were* God.

The well-known Pythagorean theorem says that, in any right triangle, the square of the length of the hypotenuse— the long side—is equal to the sum of the squares of the lengths of the other two sides. It is believed that the Pythagoreans were the first to prove this theorem and it is for this result that the name "Pythagoras" is best remembered. Paradoxically, it is this theorem which caused the Pythagoreans the most discomfort.

Consider a right triangle having its two short sides of equal length. Let's suppose, for simplicity, that each side is exactly 1-unit long. The Pythagorean theorems tells us that the square of the length of the hypotenuse is equal to 2. Consequently, the actual length of the hypotenuse is some number whose square is 2. Since the Pythagoreans believed that everything was made up of the natural numbers or of their ratios, it followed that there must be a natural number or a specific ratio of two natural numbers whose square is exactly 2. But it is clear that there is no natural number whose square is 2 since the square of 1 is 1 and the square of any of the numbers 2, 3, 4, . . . is larger than 2. Therefore, there must exist a pair of natural numbers with the property that their ratio, when squared, is equal to 2. If we call these natural numbers a and b then we are saying that $(a/b)^2$ = 2. (For typesetting reasons the fraction $\frac{a}{b}$ is often printed as a/b. The two symbols have identical meanings.)

We could give the same argument by beginning with a square having each side 1-unit long. If we then draw a diagonal we divide the square into two right triangles with short sides of length 1. The Pythagorean theorem then tells us that the diagonal of the square is a number whose square is 2. (Note the word "square" is used twice in the preceding sentence and with two different meanings. Once it refers to a geometric object having four equal sides and once it refers to the operation of multiplying a number by itself. This practice of allowing a single word to have multiple meanings is allowable in mathematics only when—as the mathematicians say—"there is no possibility for confusion." Unfortunately, what is clear to a mathematician is not always transparent to the rest of us.)

So, the *mathematics* of the Pythagoreans led them to the conclusion that the diagonal of the square was a number which, when multiplied by itself, equals 2. Their *religion* told them this number must be the ratio of two natural numbers. The mathematics was correct; their religion was wrong. For the Pythagoreans were also able to prove, using their mathematics, that it is impossible for the square of *any* ratio of natural numbers to be equal to 2. But once they had proved this, they were devastated and humiliated because it contradicted their religious beliefs. Their initial reaction was to suppress the result.

The proof that the square of any ratio of natural numbers can never equal 2 is easy and we will return to it when we discuss "aesthetics." (A mathematician often—whether lecturing or writing—will omit proofs to certain theorems.

Invariably, he says the proof is "easy" or else he says it is "obvious." By "easy" he means that the listener or the reader can do the proof without help. By "obvious" the mathematician means that it is obvious to him. The first method of avoiding the proof is the method of *intimidation*. The latter may be called the method of *higher authority*.)

In modern terminology, numbers which are the ratios of natural numbers such as 3/4 (or their negatives or the number zero) are called "rational numbers." In this terminology the Pythagorean dilemma takes the form: "There is no rational number whose square is 2" or, alternately, "The square root of 2 is not rational." The Pythagoreans liked to speak of lengths of lines and they put it this way: "The length of the diagonal of a square is incommensurable with the length of its sides." And by this they meant that, no matter how finely you divide the length of the side of a square, the diagonal cannot be measured exactly in multiples of that division. It is easy to see that the incommensurability statement and the statement about the square root of 2 are equivalent. Either interpretation is sufficient to have destroyed the religious creed of the Pythagoreans and to convince them that, if mathematics were to have any value, it had to be enriched to include numbers more complicated than those obtained by using the natural numbers and their ratios.

And it has been enriched. Mathematics now includes negative, as well as positive, numbers. It also includes numbers like the square root of 2 which are called "irrational numbers." Moreover, there are transcendental numbers and algebraic numbers and numbers which are more "unreal" than any of these and are called "imaginary numbers." In order to appreciate mathematics one must have at least an intuitive notion of these "higher" numbers and some idea of their development. We begin our development as Kronecker would have us begin, with the natural numbers.

AXIOMS

One of the things that mathematicians know and the rest of us do not is that *all of mathematics follows inevitably from a small collection of fundamental rules.* These rules are called axioms and there are several sets of axioms from which you can begin the development. In their practice of mathematics, mathematicians rarely consider the axioms. Neither do they routinely reproduce the various steps that lead from the axioms to the finished product. The typical mathematician begins his research far from the foundations of the subject and if he is successful ends it even further away. Nevertheless, each mathematician knows the axioms exist and that it is possible to start with them and to re-create, step-by-step, with logic and precision, the complete development of mathematics from the beginning right up to the contents of the most recent research paper on linear topological spaces. Moreover, each professional mathematician believes that, given pencil and paper and sufficient time, he can do this task himself—in a locked room, without help.

It is this knowledge—as much as anything else—which gives the professional mathematician confidence to deal day-by-day with a subject that scares the rest of humanity half to death. And the knowledge that the subject can be developed in this way gives the mathematician an overall view of mathematics that none of the rest of us has seen. Here's what Bertrand Russell said of this view[2]:

> The discovery that all of mathematics follows inevitably from a small collection of fundamental laws is one which immeasur-

ably enhances the intellectual beauty of the whole; to those who have been oppressed by the fragmentary and incomplete nature of most chains of deduction this discovery comes with all the overwhelming force of a revelation; like a palace emerging from the autumn mist as the traveller ascends an Italian hillside, the stately storeys of the mathematical edifice appear in due order and proportion, with a new perfection in every part.

Indeed. But only the mathematicians see the palace. No one else can penetrate the mist. Neither the humanists, who have been force-fed mathematics until they hate it, nor the scientists, who climb mathematical hillsides each workday like Sherpas, are allowed more than a fragmentary and incomplete view. Lancelot sinned greatly but still was permitted to see the Holy Grail. But you can be the greatest knight in the world and still not be allowed a glimpse of the Sangreal of mathematics. Unless, of course, you are a mathematician. The "intellectual beauty of the whole" of mathematics has remained invisible to all except the mathematicians. This invisibility provides yet another manifestation of the failure of mathematics education.

Russell's remark quoted above appears in an article, "The Study of Mathematics," written in 1902. In 1931 Kurt Gödel[3] published his famous theorems on the "inconsistency" and the "incompleteness" of mathematics. These results say essentially that the consistency of mathematics, i.e., the absence of contradictions, cannot be proved using mathematical logic. Moreover, there exist within mathematics statements which can be neither proved nor disproved. Gödel's theorems struck hard at the foundations of mathematics and were particularly discouraging to those who, like Russell, had worked hard to demonstrate that the laws of logic would ultimately provide a proof of the consistency of all mathematics. According to Harvard's Michael Guillen,[4] Russell became "disillusioned" since his own work, from the late nineteenth century onward, was based on the hope, if not the conviction, that certainty in mathematics was provable.

Nevertheless, Russell's 1902 remark remains valid in the sense that any *existing* mathematical result can be traced backward to, and derived from, a small set of axioms. Gödel's incompleteness result says it is possible to formulate mathematical statements which can be neither proved nor disproved. It says nothing about theorems for which the proof is known.

Gödel's landmark results appropriately changed the view many mathematicians had of the foundations of mathematics. But rather too much has been made of them in articles and books written for nonmathematicians. Morris Kline,[5] for example, describes them as "shattering." Ian Stewart says Gödel "threw a spanner in the works"[6] and Michael Guillen claims they "displaced logic from the center of the mathematician's world."[7]

All of this is correct. But, what for a pure logician like Russell represents a loss of "splendid certainty," represents only an esoteric curiosity to a practicing mathematician. The typical practicing mathematician worries about specific theorems, not about the probability or consistency of *all* possible theorems. The practicing mathematician normally continues, undevastated and unresigned, to prove theorems starting from some given set of fundamental laws. The logicians worry about the nonprovability of all possible theorems. The working mathematician deals with specific theorems whose proofs he can produce.

One of the sets of "fundamental laws" from which the number system can be developed is called the Peano Axioms after the mathematician Guiseppe Peano (1858–1932). Professor Peano begins by postulating the existence of a set of objects which he calls natural numbers. The objects themselves are undefined but he assumes they satisfy five axioms. The axioms are:

Axiom A: The given set is not empty. It contains an object called 1.

Axiom B: For each natural number there is exactly one other natural number called its successor.

Axiom C: There exists no natural number whose successor is 1.

Axiom D: If the successors of two natural numbers are equal than the natural numbers are equal.

Axiom E: If any collection of natural numbers has the property that it contains 1 and the property that, whenever it contains a particular natural number, it contains the successor of that natural number, then the collection actually contains all natural numbers.

Several points need to be made about these axioms. In the first place they are, except for the last axiom, simple and nontechnical. They are here stated in words, but Peano stated them symbolically. And when they are stated this way it becomes clear that only three things are being described by them. These are the ideas of *natural number,* of 1, and of *successor.* The axioms then give five rules which these notions must satisfy. From these simple rules flow all of the numbers needed for mathematical analysis.

We need not bother with details. But we need to understand the spirit of the development of the numbers which follows from the axioms; for this is precisely the spirit which defines pure mathematics. Let's turn to Bertrand Russell once more. In *Mysticism and Logic* he wrote: "Pure mathematics consists entirely of assertions to the effect that if, such and such a proposition is ture of *anything,* then such and such another proposition is true of that thing."[8] Thus, pure mathematics consists of assertions that some proposition implies some other proposition. Mathematics is the study of assertions of the form

$$p \text{ implies } q$$

where p and q are each statements about objects that live in the mathematical world of Figure 1. (A nonmathematical example is the implication "Felix is a cat implies Felix is an animal.") Pure mathematics is not concerned with the validity of either statement p or statement q but only with the validity of the assertion that "if p is true then q is true." Russell says that questions about the truth of p or the truth of q belong to *applied* mathematics. He is correct and I will return to this point later.

But we are here talking about pure mathematics and it is not possible to overemphasize the importance of the notion that the subject consists of implications of the form "p implies q." Whether p itself is true or false is not an issue. (Obviously, if the implication has already been proved, then q will be true whenever p is true. This is exactly what *proof* means.) When a mathematician proves, in a freshman calculus class, that all differentiable functions are continuous, he is not claiming that any particular function is differentiable or, in fact, even that such functions exist. Instead, he is claiming that, *if* a given function is differentiable, *then* that particular function must also be continuous.

If you want to understand pure mathematics then you must first understand this point about implications. All pure mathematicians have this understanding as part of the fibre of their being. It is as natural to them as the air they breathe. *And this understanding is one of the three clearly defined characteristics which separate those people who understand mathematics from those who do not.*

The development of the number system from the Peano Axioms will then consist of the formulation of statements of the form "p implies q" and the proving of each statement. As you get further from the axioms you must introduce definitions of new objects in order to make appropriate assertions. But each new definition must be consistent with the original axioms and the proof of each new assertion must use *only* the axioms and the assertions which have already been proved. Therefore, at each stage of the development, you can use only what you have specifically assumed as axiomatic or what you have already proved from the axioms. The word "only" is essential. *And the understanding that it is essential is the second fundamental characteristic mathematicians possess which distinguishes them from others.*

Suppose, for example, that you are teaching the development of the number system from the Peano Axioms and you have just written the axioms on the blackboard. Just as you finish writing axiom E a student in the front row asks: "What, professor, does all this tell us about 2 plus 2?" How do you respond?

The answer is (a mathematician would say "the answer, *of course,* is") that the student's question is meaningless. Because at this stage all we know about the natural numbers is what the axioms tell us and they tell us nothing about addition. In fact, at this stage it makes no sense even to talk about "2." All we know from the axioms is the name of a single natural number, namely, the number 1. Before we can discuss "2" we must define what this symbol means.

Once, as dean, I gave a talk to a group of social scientists. The man who introduced me wanted to make a joke about mathematics and mathematicians. So he told the assemblage a slightly revised version of a tired old joke. The joke goes like this. You ask the question "What is 2 plus 2?" and the answer you get depends on the occupation of the person asked. You ask your accountant and he responds, "What do you want it to be?" The president of the United

States says: "2 plus 2 is four billion." Your psychiatrist says: "Lie down on my couch and tell me why you want to know."

As I say, it is a tired old joke. But the introducer told it at the postcocktail hour and both he and the audience found it amusing. Then he said:

> However, our speaker is a *mathematician*. And if you ask a mathematician "What is 2 plus 2?" he will say that he doesn't know. But he does know that, *if* 1 plus 1 equals 2, *then* 2 plus 2 equals 4.

The introducer giggled and the audience howled. I walked to the podium straight-faced and said that the joke was not a joke at all but was exactly the right answer to the question. And it is. Every mathematician knows.

You will notice I have labelled the Peano Axioms with the letters A through E rather than with the integers 1 through 5 and I have done this precisely because, in the beginning, none of the numbers after 1 has a name. In fact, the general notion of "after" has not even been established. What we do know is that, by Axiom A, there is a number called 1. And by Axiom B we know that each number has a unique successor. We also know by Axiom C that the successor of 1 is not 1; that is, the successor of 1 is some number other than 1. We can give that number a name if we wish. And we do wish; we call the successor of 1 by the name 2.

It is natural to attempt to proceed in this manner. We know, again by Axiom B, that there is a successor to the number 2. It is tempting to call it by the name 3. But we can't be so hasty. Because the bright student in the front row will immediately ask us how we know that the successor of 2 is not 2 itself. After all, the axioms say nothing about a number called 2. So we must first prove the following:

Theorem: The successor of any natural number is different from the number.

The proof of this theorem is not difficult but when you
try it you see you need a preliminary result. Namely, you
need:

Theorem: Different natural numbers have different
successors.

Therefore, the latter result is first proved and *then* the
former theorem is established.

I will not give the proofs of either theorem here because they fall outside the purpose of this book. I want you
only to see the necessity for the first theorem. You cannot
simply give the successor of 2 a new name until you first
establish that it is different from 2. This is a good example
of what I called the second characteristic of people who
understand mathematics. Namely, that you *know,* at any
stage in the development of mathematics, only what is
given by the axioms or what you have already *proved.*

As for the first characteristic—pure mathematics consists of assertions of the form "*p* implies *q*"—note that the
first theorem may be written as

Theorem: If y is the successor of x then y is not equal to x.

This is now clearly an assertion of the form "*p* implies *q*"
where p is the statement "y is the successor of x" and q is
the statement "y is not equal to x."

I have not yet mentioned the third characteristic
shared by those who truly comprehend mathematics. This
characteristic is the understanding of the need for *complete
precision* in the statement of the axioms, the definitions,

and the proofs of the assertions. The goal of mathematics is absolute—and not approximate—truth. And there can be no certainty about the truth of mathematical results without the presence of total precision in the procedures leading to these results. If, for example, you want to prove that the product of two odd numbers is always odd (assuming that the notions of "odd" and of "product" have been properly defined), then you must prove this for *any* pair of odd numbers. It is not sufficient to observe that the result holds for 3 × 5 and 7 × 9 and 65 × 45 or, indeed, that it holds for any pair of odd numbers which occur to you. You must prove the result for *arbitrary* pairs of odd numbers. And each step in the proof must be clear and precise and must follow from the axioms or from other known results.

Albert Einstein said of mathematics[9]:

> One reason why mathematics enjoys special esteem, above all other sciences, is that its laws are absolutely certain and indisputable, while those of all other sciences are to some extent debatable . . .

The certainty of the laws (proved theorems) of mathematics holds only because of the complete precision involved in the derivation of those laws. In fact, without this certainty pure mathematics would be reduced from special esteem to little esteem. For the objects of pure mathematics live only in the mathematical world. They are abstract, and not real, objects. *Knowledge* about objects which live only in the mind would be of little value unless we can be sure that what we know is the truth. And truth which comes from deduction and not from observation is possible only by way of complete precision. Pure mathematics is precise or else it is nothing.

Incidentally, the totally abstract nature of pure mathematics is often put forward as an explanation for the high degree of difficulty the subject presents for all but a small subset of the population. This "explanation" comes from both mathematicians and from nonmathematicians alike. "Only a gifted few," the saying goes, "have the capacity to think abstractly." This is a convenient explanation and there is something in it for everybody. The mathematicians like it because it gives them an intellectual patina they do not have to earn (since they belong to the gifted few) while simultaneously providing them an excuse for their failure to transmit their knowledge of mathematics to outsiders. And the outsiders find it comforting to believe that their inability to understand mathematics results from a handicap beyond their control and not from simple failure of nerve or self-discipline. The "explanation" supplies a ready-made reductive explanation for failure at both ends of the mathematical spectrum. Unfortunately, the explanation is false.

Abstract thinking is nothing more than *selective* thinking. What you do when you create an abstract model of a real object is to select certain features of the object on which you focus your attention. You can form a variety of abstractions of the same object depending on your needs or interests. To the girl next door, for example, the abstract model of a young man may be a complicated mental image of certain of his characteristics such as his smile and his personality while, to a college football recruiter, the same young man is modeled by a set of statistics describing his performance as a running back in his high school senior year. The two models are very different but both are meaningful. They simply reflect different modes of selective thinking.

Abstract thinking is commonplace. And so are abstract objects. The Princeton mathematician Salomon Bochner puts it this way[10]:

A cookbook is abstract, and so are the Yellow Pages in a telephone directory. Any academic textbook, on whatever level

and in whatever subject, must be abstract in a more than per-
functory sense or else it is useless.

Exactly. It is not abstraction which makes mathematics difficult. Rather, it is precision. Mathematics is difficult
because, unlike any other discipline, it demands complete
precision.

You cannot fake. In mathematics, no one can be
fooled. You can either prove differentiable functions are
continuous or you cannot. Not the remotest possibility exists that you can bluff your way through it. Abstraction is
commonplace. Precision is unnatural and hard.

But I am getting ahead of our story. We are now at
the stage where we have set down the Peano Axioms and
we have the names of two natural numbers, 1 and 2.
What's next?

We could proceed to name the natural numbers one by
one by going from each number to its successor as we did in
going from 1 to 2. Let's assume in fact that this has been
done in the usual manner so that the successor of 2 is called
3, the successor of 3 is called 4, and so on. This gives us the
natural numbers named in the familiar way but there is
little we can do with them since we have defined no operations allowing us to make combinations and we have no
notion of order. We cannot, for example, say "4 is greater
than 2" because we have not defined the notion of "greater
than." (Remember, all we can use are the axioms and what
has been proved from the axioms.) Similarly, we cannot
talk about "2 plus 3." We need first to define appropriate
operations on the natural numbers and a notion of
ordering.

The details of the formulation of definitions and the
subsequent development are not essential for our purposes
and I will give only the barest of outlines. We need only

understand that the process exists and that it is the prototype of all work in pure mathematics, and we need some
familiarity with the higher numbers the process yields.

THE INTEGERS

The definition of *addition* comes next. It is easy, and
natural, to define $x + 1$ where x is any natural number. You
define it to be the successor of x. This gives us the string of
results $1 + 1 = 2$, $2 + 1 = 3$, $3 + 1 = 4$, and so on, since we
have names for these successors. It is more complicated to
define $x + y$ where x and y are *any* two natural numbers
since the definition involves the use of Axiom E, the most
complicated of the Peano Axioms. (Axiom E is called the
induction axiom and has the following physical analog.
Consider a row of dominos standing on edge. Suppose they
are placed so that, whenever any particular domino falls,
the domino after it falls. Now push the first domino over.
Axiom E says that *all* the dominos fall.)
Let's assume that addition has been properly defined.
Next, we *adjoin* to the natural numbers an element having
the property that, when it is added to a number, the result is
the number itself. This new element, denoted by 0, is the
number zero. Zero is not a natural number but is something
extra, a new number having the properties $0 + 0 = 0$ and x
$+ 0 = x$ for each natural number x. A fancy name for zero is
"additive identity."
We notice that we still cannot do simple arithmetic.
For example, we cannot solve the equation $x + 2 = 1$. There
is simply no natural number x which, when added to 2, will
produce the number 1. In order to do this we need more
numbers. So we introduce (through proper definitions consistent with the axioms) the negatives of the numbers we

already have. This means that for each natural number x we introduce a new number y having the property that $x + y = 0$. The number y is called the *negative* of x. Once this has been accomplished, we have the set of numbers . . . , -4, $-3, -2, -1, 0, 1, 2, 3, 4, \ldots$.

The numbers written in this infinite list are called the *integers*. (The solution of $x + 2 = 1$ is now $x = -1$.)

Next, we define an operation called *multiplication* in such a way that it is meaningful to talk about the *product* of any two integers so that the new operation has the desired relationship with the notion of addition. In particular, we define multiplication in such a way that it is *distributive* over addition. [To say that multiplication is distributive means that $x(y + z) = xy + xz$ for all integers x, y, and z.]

And somewhere in this "definition–theorem–proof" process we introduce the notation of "ordering" so that it is meaningful to talk about one integer being greater than or less than another integer. Once this is done we have on hand the collection of positive and negative integers on which we have laid the operations of addition and subtraction and the notion of order. One can do a lot with this set of mathematical objects but not nearly enough. We cannot, for example, *divide* one integer by another.

THE RATIONAL NUMBERS

At this stage, we have infinitely many integers with which to work and we can do a fair amount of arithmetic with them. At our own disposal are the natural numbers 1, 2, 3, . . . , their negatives $-1, -2, -3, \ldots$, and the number 0. If the economy worked only with whole dollars these numbers (the integers) would be sufficient for banking purposes. A bank statement bottom line of 0 says your account contains no funds, a balance of 1,000 means you have one thousand dollars, and a $-1,000$ balance says you are a thousand dollars overdrawn. But banks deal with fractions of dollars, with decimals. Banks *divide* whole dollars into pieces.

But at this stage we cannot divide one integer by another. We do not yet have *fractions*. Nor can we solve even simple equations.

For example, we cannot solve the simple equation $2x = 3$. That is, there is no *integer x* in our infinite collection of integers which, when multiplied by 2, will give the number 3. Evidently, we need more numbers. So, we next introduce the notion of the *ratio* of two integers, i.e., we define new numbers which are of the form x/y where x and y are integers. (Recall that x/y means $\frac{x}{y}$.) The exact process by which this can be done is reasonably straightforward but it involves too much technical detail to allow presentation here. What needs to be understood is that the process leads to new numbers which formally are *ratios* of integers and, consequently, are called rational numbers. The rational numbers, therefore, are just the ordinary fractions which have integers as both numerator and denominator. Examples of rational numbers are $3/2$, $-17/5$, and $6/1$. (The solution of $2x = 3$ is $x = 3/2$.)

Next we extend the definitions of multiplication, addition, subtraction, and the notion of ordering to the rational numbers. And we do it in such a way that each *new* number of the form $x/1$ can be simply identified with the integer x. When this has all been accomplished we have a much richer collection of numbers for doing arithmetic and now we have the new notion of *division*. We think of the rational number 3/2, for example, as the number obtained when 3 is divided by 2.

Incidentally, it turns out that, when you form the rational number x/y, you must always have y *not* equal to 0. That is, division by zero is not allowed. The basic reason for this restriction is that one of the theorems which is proved in the careful development of the rational numbers (which is only outlined here) says: For any rational number z it follows that $0 \cdot z = 0$; that is, the product of any rational number and zero equals zero. A second theorem states: $x/y = z$ if and only if $x = yz$. (Here, we are following the common practice of using both the dot and the juxtaposition of symbols to mean multiplication. Thus, xy and $x \cdot y$ both stand for the product of x and y.) This second theorem says that if the rational number x/y is equal to another number, denoted by the single symbol z, then it must be true that the numerator of x/y equals the denominator multiplied by z. For example, $6/3 = 2$ because $6 = 3 \cdot 2$.

Now suppose we could assign meaning to $x/0$. Then $x/0$ would be the name of some rational number z. Thus, $x/0 = z$. So, by the second theorem, with 0 playing the role of y, we would have $x = 0 \cdot z$. But by the second theorem, zero multiplied by any number equals zero. Hence, $x = 0$. Therefore, $x/0$ can have meaning only if $x = 0$, which tells us that, if the denomination of a fraction were allowed to be zero, then the numerator must also equal zero.

So, our original statement $x/0 = z$ turns out to be $0/0 = z$. But this assignment of value to the symbol $0/0$ is flawed because the second theorem now tells us that $0/0 = z$ can hold only if $0 = 0 \cdot z$. But the first theorem now informs us that this equation holds for *every* number z. Therefore, $0/0$ could be *any* number, i.e., $0/0$ has *indeterminant* value and is, thus, meaningless.

The only way out of this quandary is to disallow division by zero. The bottom line then is *division by zero is not a permissible operation.*

The rational numbers are exceedingly rich. With them you can do most of the mathematics you will ever need to do. The rationals are, in fact, the mathematics used by digital computers and are the mathematics used in the increasingly fashionable college courses known as "discrete mathematics." The rational numbers are the numbers that Pythagoreans believed to make up the entire universe. The rational numbers are, indeed, abundant. But not abundant enough. With them, for example, you cannot solve the equation $x^2 = 2$. It is not difficult to prove that there exists no rational number whose square is the number 2. And it was the proof of this simple theorem which threw the world of the Pythagoreans into disarray. (According to Morris Kline, the theorem is attributed to Hippasus of Metapontum, who proved it while he and the other Pythagoreans were at sea. For producing a result which denied the Pythagorean belief in the universality of rational numbers, Professor Kline says Hippasus was literally thrown overboard, providing an early example of publish *and* perish.)

The process of development of the numbers must, therefore, be continued if we are to solve routine equations and deal with simple geometric objects such as a square whose sides are 1-unit long (because such a square has a diagonal with length x where $x^2 = 2$).

THE REAL NUMBERS

So far, we have outlined the chain of development that begins with the fundamental assumptions about the natural numbers—the Peano Axioms—and proceeded to the integers and then to the rational numbers. Of course, a natural number is an integer and an integer is a rational number (for example, the integer 6 is the rational number 6/1). What we have done is *outline* the process by which we systematically enlarge our collection of numbers. The next step is to further enrich our set of numbers so that we can talk about numbers like the "square root of 2." The process here becomes subtle and difficult and we can only be descriptive. One way to think about what needs to be done is to imagine the rational numbers as corresponding to points on a horizontal line in the following manner (see Figure 2).

Fix a point on the line and mark it with the number 0. Select a convenient distance to correspond to the number 1 and mark the line in each direction from 0 with points whose distances from 0 (called the origin) are 1, 2, 3, ... The positions to the right of the origin are marked with positive integers and those to the left with negative integers. By taking fractions of these distances you can mark points on the line corresponding to each of the rational numbers. Some of these points are shown in Figure 2. Now a certain theorem about the rational numbers says that between any two of them lies another. This result is also obvious from experience and this obviousness is one of the reasons the Pythagoreans believed the rationals were all one needed to interpret the world. For example, 1/2 lives between 0 and 1. And between 1/2 and 1 you find the rational number 3/4. And 5/8 lives between 1/2 and 3/4. In general, if a and b are rational numbers and a is less than b, then the

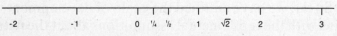

FIGURE 2. The real number line.

rational number $(a + b)/2$, which is the average of a and b, lives between a and b.

It follows easily from these considerations that the rational numbers are "dense" on the line of Figure 2. That is, in any interval of this line you can find a rational number and, in fact, infinitely many rationals. But there are still "holes" in the line. In particular, there is no rational position x on the line which corresponds to the distance having the property that $x^2 = 2$. Or, to put it another way, there exists no rational number that sits at a distance "the square root of 2" either to the left or to the right of the origin. To be sure, there are rational positions on the line arbitrarily close to this position. For example $(1.4)^2 = 1.96$, which is less than 2, and $(1.5)^2 = 2.25$, which is greater. Hence, the rational number 1.4 sits to the left of the square root of 2 and 1.5 to the right. (Note that 1.4 and 1.5 are rational since they are, respectively, 14/10 and 15/10.) Similarly, 1.41 and 1.42 live, even closer, to the left and the right, respectively, of the square root of two.

The objective of the development process is then to "fill in" the holes in the line so that, in particular, there exists a number corresponding to the square root of 2. (There are many other holes that must be filled. For example, it can be shown that there are no rational numbers satisfying $x^2 = y$ where y is any of the numbers 2, 3, 5, 7, 11, 13. Or, in fact, where y is an infinity of other numbers. To "complete" the line, you must fill in an *infinity* of holes.) It is important to understand that the geometric discussion we are following is only heuristic and the logical development must proceed from the axioms and the previous theorems without geometric considerations. That is, the proofs must not depend on sketches or pictorial representations but only on the axioms and prior theorems. In fact, at this stage, you cannot actually speak of "the square root of 2" since no such number exists in the rationals. The exact purpose of the (omitted) development is to produce, in a logical manner, this, and other, new numbers.

But the geometric interpretation is helpful since it shows there are rational numbers arbitrarily close to the number we want to define. The actual procedure exploits this observation and "adjoins" to our set of rational numbers new numbers which correspond in a technical sense to limits of appropriate sequences of rational numbers. ("Limits" and "sequences" are notions which are precisely defined in the actual "filling in the holes" development.) When this has been done, and the notions of addition, subtraction, multiplication, division, and ordering have been defined for these new objects we have an extended system of numbers which contains, in particular, the solution to the equation $x^2 = 2$. Now we can talk about "the square root of 2" and invent a symbol for it: $\sqrt{2}$. But the process which gives meaning to this symbol is not only subtle but slow. And it was not until the time of Dedekind (1831–1916) that precise sense was made of it. (Far too late to bring comfort to the Pythagoreans or to pull poor Hippasus from the sea.)

This new system represents the "completion" of the line of Figure 2 and it is called the set of *real numbers*. The line, consequently, now becomes known as the real number line, or simply *the real line*. The real numbers now contain the rational numbers as a proper subset. Real numbers which are not rational, such as $\sqrt{2}$, are called *irrational* numbers. There are *no* holes in the real line. The holes which existed before have now been filled in with the irrationals.

It might be helpful to think of the development so far as something like this: the *natural numbers* are the ordinary counting numbers 1, 2, 3, ... The *integers* are the counting numbers together with their negatives and the number zero. The *rational numbers* are ratios of integers such as 3/2, 22/7, and −5/16, except that no rational number can have zero as its denominator. The *real numbers* are all of these together with any new numbers like $\sqrt{2}$ that can be obtained by forming algebraic combinations and solving certain equations. A rule of thumb asserts: Any number appearing in mathematics courses up to and including calculus is a real number. (An important exception to the rule occurs in elementary algebra where students are asked to solve quadratic equations and learn that under some conditions no real-number solutions exist.)

We should mention certain properties of these numbers which, in a rigorous development, would be proved as theorems. First, any real number has a decimal expansion representation. Thus, any real number may be written as an integer followed by a decimal point and a string of integers. For example, 3/2 = 1.5. The string of digits following the decimal may be infinite as, for example, 1/3 = 0.3333333. . . . In fact, another theorem asserts that any irrational number is represented by an infinite decimal where the string of digits neither terminates nor has a repeating pattern. For example, $\sqrt{2}$ = 1.41421356 . . . and there is neither end nor a repeating pattern to the digits after the decimal. Conversely, any decimal represents some real number and the rational numbers are exactly those numbers corresponding to terminating or to repeating decimals. Hence, the infinite decimal .123456789101112 . . . represents some real number—evidently between 0 and 1. And, since the decimal is infinite and nonrepeating, the number it represents is not a rational number. This number, which is as "natural" as the counting numbers are natural, was studied by the mathematician Kurt Maher, who died in 1988. Maher proved that this number belongs to a class of real numbers called *transcendental numbers*.

An expression like $ax^2 + bx + c$ or $dx^5 + gx^3 + hx^2 + j$, i.e., a sum of powers of x multiplied by real numbers called "coefficients" (the numbers a, b, c, and d, g, h, j, in the examples), is known as a *polynomial*. Polynomials are often indicated by symbols such as $p(x)$ which is read "p of x"; $p(x)$ is a real number which depends on x. If $p(x) = 0$, then the number x is called a *zero* of the polynomial. For example, $x = 2$ is a zero of $p(x) = x^2 - 4$ since $p(2) = 2^2 - 4$, and $4 - 4 = 0$.

If $p(x)$ denotes a polynomial which has only *integer* coefficients (such as $p(x) = 3x^2 + 2x + 6$) and if y satisfies the equation $p(y) = 0$, then the real number y is called an *algebraic* number. Thus, an algebraic number is a real number which is a zero of a polynomial with integer coefficients. The square root of 2 is algebraic because it is a zero of the polynomial $x^2 - 2$ (because $x^2 - 2 = 0$ if and only if $x = \sqrt{2}$ or $x = -\sqrt{2}$). Each rational number a/b, is algebraic since it satisfies the equation $bx - a = 0$. Numbers which are *not* algebraic are called *transcendental*.

At first glance these numbers seem strange and rare. Yet it is surprisingly *easy* to prove the existence of many, many transcendental numbers. And it is surprisingly difficult to prove that any *particular* number is transcendental. The number π—the ratio of the circumference of a circle to its diameter—is transcendental. Another famous transcendental number is the number denoted by e which, although it is often called the "base of the natural logarithm," plays a role in mathematics far deeper than anything most people can imagine from their memory of what they studied in high school having to do with "logarithms." This number is, in fact, related to π in many deep and beautiful ways. It is by no means obvious that either π or e is transcendental. (The number e, by the way, lives between 2 and 3 and has the decimal expansion $e = 2.71828459045. . . .$)

Mathematicians develop a "feel" for transcendental numbers from their experience and insight. Ask any one of them, for instance, if the number $e^{\pi\sqrt{163}}$ (the number e raised to the "power" $\pi\sqrt{163}$) is an integer and he will immediately say "no." To a mathematician, this number lives worlds away from any integer. Ask the same question of a scientist and he will pull out his calculator and "compute" the number. And how he answers the question depends on the number of significant digits held in his machine, because $e^{\pi\sqrt{163}}$ differs from an integer only in the twelfth decimal place which might, or might not, show up on a pocket calculator display.

Our number is actually transcendental and this can be shown by appeal to the 1932 Gelfond–Schneider Theorem which says that, if a is an algebraic number which is different from 0 or 1, and if b is an algebraic irrational number, then ab is transcendental. In order to apply this theorem it is necessary to rewrite $e^{\pi\sqrt{163}}$ in a manner possible only through the use of numbers which are even more abstract and more general than are the real numbers. To find these new numbers and to complete our journey from the natural numbers and the Peano Axioms, we need to discuss yet one more enrichment of our number system. We need to go all the way to the complex numbers.

THE COMPLEX NUMBERS

One of the facts about the real numbers which puzzles elementary students is that the product of two negative numbers, like the product of two positive numbers, is positive. The associated result which asserts that the product of a positive number and a negative number is negative causes, at least for the natural numbers, little confusion. The student can, for example, think of $(-10)(5)$ as being the same as $(5)(-10)$ and he can interpret this as meaning -10 added to itself 5 times. And he has no trouble seeing that the sum of five -10s is exactly -50. Similarly, the student has no difficulty with the fact that the product of two positive numbers is positive. (Note that I am once again indicating multiplication by juxtaposition of symbols. Thus, "ab" means the product of "a" and "b." When signs are involved, or the expressions a or b are themselves complicated, I will indicate this multiplication by (a) (b). Sometimes, I will denote a times b by $a \cdot b$, where the dot denotes the multiplication sign. I will avoid, as much as possible, the use of "$\times$" as a symbol for multiplication. Mathematicians almost never write $a \times b$ for "a times b" unless they are talking to nonmathematicians and then they use the symbol with condenscension.)

All students of elementary mathematics, however, are puzzled by the fact that $(-5)(-10) = 50$. How, they ask, can this be? It makes no sense to think of adding -5 to itself -10 times. Nor can one add -10 to itself -5 times. It is possible to provide some heuristic "arguments" for this phenomenon but they are mostly unconvincing and mainly serve only for the amusement of the mathematician who invents them.

One such argument is given by Michael Guillen in his otherwise fine book, *Bridges to Infinity*. Guillen asks his reader to consider an election in which a ballot cast for a particular proposition is worth 10 points while a ballot cast against it is worth -10 points. He describes a "negative voter" as a person who is qualified to vote but who does not. And he then observes that if five people vote for the proposition (five "positive voters") then it gains a total of $(5)(10) = 50$ positive points. Then he says:[11]

> If five negative voters who are against the proposition (-5) all neglect to cast their con ballots (at -10 points each), then in this way too the proposition has gained, indirectly, a total of fifty (-5×-10) positive points. It is a fifty point advantage that the proposition would otherwise not have had.

I will not bother to criticize this "argument" except to point out that the notation used in the quote is Guillen's and not mine, and to suggest that anyone who finds this discussion helpful is welcome to it. Guillen fails evidently to convince himself since he writes two sentences later: "This rule is often called the law of signs, and of all the algebraic rules for combining negative numbers it seems to perplex people the most."[12]

The law of signs is actually easy to comprehend provided you were taught mathematics correctly by a teacher who understood the subject, and provided you were taught by someone who believed in teaching, and not in patronizing, the students. For, at whatever level you were introduced to the real numbers, whether in high school or in college, you learned you can manipulate these numbers in certain ways and you learned that $z \cdot 0 = 0 \cdot z = 0$, for any real number z. Moreover, you learned that $w + (-w) = 0$, for any number w. (In a careful development, these facts will be theorems and will have been proved using previous theorems and the Peano Axioms and not just *learned* by authoritative communication from teacher to student.) Nothing more is needed to establish the law of signs: Let a and b be any two real numbers. Consider the number x defined by

$$x = ab + (-a)(b) + (-a)(-b).$$

We can write

$$x = ab + (-a)[(b) + (-b)]$$
$$= ab + (-a)(0)$$
$$= ab + 0$$
$$= ab.$$

Also,

$$x = [a + (-a)]b + (-a)(-b)$$
$$= 0 \cdot b + (-a)(-b)$$
$$= 0 + (-a)(-b)$$
$$= (-a)(-b).$$

So we have

$$x = ab$$

and

$$x = (-a)(-b).$$

Hence,

$$ab = (-a)(-b).$$

That's all there is to it. Q.E.D., as the mathematicians say. (*Quod erat demonstrandum,* or "quite easily done"—you choose.)

The preceding argument has great advantages over the "explanation" offered by Guillen. First, the argument contains mathematics and not just air. Second, it lives close to the truth. (The argument is "true" if the results needed [such as $z + 0 = z$] to do the simple manipulation have been established.) Moreover, the argument is nonpatronizing. The perplexity either goes away or else it remains because you fail to understand "mathematics," not because you cannot comprehend the association of an artificial tale about "negative voters" with what it means to multiply, say, $-e$ by π.

One of the consequences of the law of signs is that, if x is any real number other than zero, then x^2 is always positive. Consequently, there exists no real number whose square is negative. In particular, there is no real number whose square is -1. Thus, if you have only the numbers that live on the real line you cannot—in spite of all the infinity of numbers at your disposal and the vast richness of structure they possess—solve the elementary equation $x^2 = -1$. To do this, you need numbers of yet another type. You need $\sqrt{-1}$—the square root of -1.

In a logical development you consider a new set of objects consisting of pairs of real numbers. Thus, you look at the collection of all pairs (x, y) where x and y are real numbers. Upon this collection you define, in an appropriate manner, addition and subtraction. Once this has been done, you prove theorems about this new mathematical object in a manner analogous to that used in the creation of theorems about the real numbers. This new set, together with the defined operations, is called the set of *complex numbers.*

Therefore, the complex numbers consist of the set of all pairs of real numbers where we have (presumably) defined a precise way of adding and multiplying any two of these pairs.

The first observation to be made about the complex

numbers is that, since each complex number is a pair (x, y) of real numbers, these new numbers correspond one to one to points in a plane. (The one-to-one correspondence arises from the observation that any point p in the plane is uniquely described by two coordinates (x, y) which measure its distance from two lines called the x-axis and the y-axis.) The plane we use is the familiar plane of analytic geometry (see Figure 3), except we now have a richer structure because complex numbers can be multiplied and added in addition to being plotted as points. Points on the x-axis all have coordinates of the form $(x, 0)$, and it is possible—in a precise, mathematical manner—to associate $(x, 0)$ with the real number x so that the x-axis of the complex plane becomes a copy of the real line, because each point on the x-axis has y coordinate equal to zero and, therefore, is represented by a pair of the form $(x, 0)$. With this association, the real numbers become a subset (the x-axis) of the complex numbers.

So far, we have not dealt with the precise notion of addition or multiplication of complex numbers. These *definitions* are, respectively,

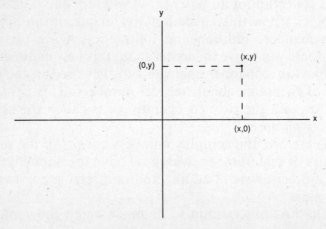

FIGURE 3. The complex plane.

$$(x_1, y_1) + (x_2, y_2) = (x_1 + x_2, y_1 + y_2)$$

and

$$(x_1, y_1) \cdot (x_2, y_2) = (x_1 x_2 - y_1 y_2, x_2 y_1 + x_1 y_2).$$

(Notice that the operations which are being defined are on the left side of these equations. The operations on the right are the ordinary operations of addition and multiplication of real numbers.)

Points on the y-axis each have coordinates of the form $(0, y)$. It turns out (i.e., it is a *theorem*) that the complex number $(0, y)$, when multiplied by itself, becomes the complex number $(-y^2, 0)$. That is,

$$(0, y)(0, y) = (-y^2, 0).$$

This comes out of the above definition of multiplication of complex numbers if we set $(x_1, y_1) = (x_2, y_2) = (0, y)$. If we set $y = 1$ in this equation, we get as a special case

$$(0, 1)(0, 1) = (-1, 0).$$

If we now denote the complex number $(0, 1)$ by the letter i and use our identification of pairs of the form $(x, 0)$ with the real number x, the equation becomes

$$i \cdot i = -1.$$

Thus, we have

$$i^2 = -1$$

and we have created a number i whose square is -1. Hence, $i = \sqrt{-1}$. [Incidentally, the use of the symbol i for the complex number $(0, 1)$ is traditional. It goes back to the early mathematical analysts who considered the equation $i^2 = -1$ so mysterious that they referred to i as an *imaginary* number. This terminology arose from the fact that the square of any *real* number is nonnegative. That is, $x^2 \geq 0$ for any real number x; for example $3^2 = 9$ and $(-3)^2 = 9$.]

Much, to be sure, has been omitted from this discussion. But the details are insignificant for our purposes. At this stage we need only be aware that a logical development of the complex numbers exists, that it is relatively straightforward, and had we time and space we could attend to all the details. (It's far easier to go from the real numbers to the complex numbers than to make our earlier step from the rationals to the reals. If the opposite seems true here it's because the difficult steps in the "earlier" development involve the notion of sequences and have been glossed over.) The development—once it has been completed—results in a new set of numbers, containing the set of real numbers, upon which we have operations of addition, subtraction, multiplication, and division. Moreover, within the complex numbers, we are able to solve equations which cannot be solved using only the real numbers.

One property which the complex numbers lack, however, is the property of order. No meaning can be assigned to statements concerning the *size* of complex numbers. Given two complex numbers, z and w, it makes no sense to say "z is larger or smaller than w"—unless, of course, z and w are real numbers in which the usual notion of "greater than" applies.

You should also notice that the development of the complex numbers from the real numbers is, as was the development of the reals from the rationals, just more mathematics. The logical creation process contains neither mystery nor magic, only mathematics. But this was not the case with the historical development. Early on, the complex numbers were shrouded in mystery and paradox.

The part of mathematics which now belongs to algebra developed significantly long before real precision was brought to mathematics. As early as the sixteenth century numbers were being manipulated with impunity before mathematicians understood exactly what was meant by the *concept* of number. These formal calculations produced startling results including, in the seventeenth century, the solving of cubic and quartic equations, i.e., in finding zeros of polynomial expressions involving third and fourth powers of real unknown variables. Mathematicians who did not fully comprehend irrational or even negative numbers did not hesitate to combine them in formal ways to create new mathematics. They did not hesitate—*except* when it came to the square root of negative numbers.

René Descartes, in about 1630, called the square root of a negative number "imaginary." His deep concern over these objects persists to this day in our use of the letter i to denote the square root of -1. Descartes believed that the occurrence of "imaginary" numbers in any formal calculation meant that the problem was, in fact, not capable of solution. According to Ian Stewart, Isaac Newton shared this opinion. Not until the nineteenth century were complex numbers made "legitimate"—mainly through the work of Gauss, who was interested not in the algebra of these numbers but rather in their use in mathematical analysis. Gauss thought of complex numbers as points in a plane and developed many of the fundamental ideas which led to the branch of mathematics now known as "complex analysis." Only in 1837 did William Rowan Hamilton formalize the geometric notions of Gauss by identifying a complex number with its coordinates (x, y) and setting down definitions of addition and multiplication in terms of these coordinates. The notation of the early analysts prevailed, however, and we continue to write a complex number, as they did, in the form $x + iy$. The (x, y) pair notation lends itself to the formal development of the complex numbers, while the $x + iy$ notation is easier to manipulate. The way to manipulate complex numbers essentially follows the rule: manipulate formally as you would with real numbers, replacing i^2, each time it occurs, with -1 [for example, $(2 + 3i)(4 + 5i) = 8 + 10i + 12i + 15i^2 = 8 + 22i + 15(-1) = 8 - 15 + 22i = -7 + 22i$].

Complex analysis became the glory of nineteenth-century mathematics. Some of the greatest mathematicians contributed to the development of this branch of mathematics. And their greatness rests largely on their contributions in this area. Gauss recognized the value of the complex integral in 1811. Cauchy proved the beautiful integral theorem which now bears his name in 1825 and showed the connection between analytic functions and a system of partial differential equations. Riemann developed the geometric theory of complex analysis and, in about 1860, stated his great conjecture concerning the location of the complex numbers at which the zeta function has the value zero. (The zeta function is a rather natural, complex-valued function defined by a kind of "infinite sum" called an infinite series. An exact description of the zeta function lies beyond the scope of this book.) This conjecture—the famous Riemann hypothesis—remains unproved. It stands today, challenging and deep, as the premier open problem in pure mathematics.

In the latter part of the nineteenth century, Karl Weierstrass brought forward his theory of power series, demolishing once and for all the mysticism of imaginary numbers by bringing full rigor to complex analysis. As the century ended, Hadamard and De La Vallee Poussin, using complex analysis, independently proved the singular result known as the "prime number theorem."

By definition, a prime number is an integer greater

than 1 which has no divisors except itself and 1. There exists only one even prime number, namely, the number 2. Any other even number, say 36, is divisible by 2 and, therefore, cannot be prime. The first few primes are 2, 3, 5, 7, 11, 13, and 17. When you think about prime numbers one of the first questions which arises is how many are there? Does there exist a largest prime or do they go on and on? A reasonable initial conjecture is that there are only finitely many; i.e., there is some largest prime number. For, if a number is sufficiently large, chances are that *something* must divide it, if only because there are so many possibilities. Take the national debt, for example. Reduce it to pennies and you have an enormous number. Now square the number. The result is a number of staggering magnitude. It, however, cannot be prime because it is a perfect square and so it is divisible by the number which was squared to produce it. But now take this number and increase it by 1. Is the new number prime? It's not clear. But it is so huge that there are an enormous number of candidates for divisors. Chances are it is not a prime number. Or is it?

Never mind. Euclid proved in 300 B.C. that the number of primes is infinite. And his argument stands today as an example of *elegant* mathematics. It goes like this:

Suppose there are only finitely many prime numbers. Let n be the number of primes. Denote this finite set of primes by p_1, p_2, p_3, . . . , p_n. Look at the number x given by

$$x = p_1 p_2 p_3 \cdots p_n + 1.$$

Now x is clearly a positive integer and it is clearly not a prime. (It is 1 larger than the product of *all* the primes and, therefore, is larger than any single prime.) So x must be divisible by some prime. (Since x is not prime, it has proper divisors. So $x = a \cdot b$ where a and b are integers larger than 1. If a or b is prime, then we have a prime that divides x. Otherwise, a and b have factors which are either prime or themselves have factors. Thus a = a_1a_2 and $b = b_1b_2$, say. Then $x = a_1a_2b_1b_2$. Continuing in this manner we conclude that x is actually a product of primes and hence is divisible by a prime.) But x cannot be divisible by any of the primes $p_1, p_2, p_3, \ldots, p_n$ because each of these, upon division into x, will clearly leave a remainder of 1. (For example, if 2, 3, 5 were *all* the primes, then $x = 2 \cdot 3 \cdot 5 + 1 = 31$ and dividing 31 by either 2, 3, or 5 leaves a remainder of 1.) Consequently, there must exist some other prime number other than those in our list. This is a *contradiction* since our list contains *all* the prime numbers. Thus, there are infinitely many prime numbers.

Euclid's proof is interesting for several reasons: it is within reach of all willing to think hard about it, no matter what their mathematical background may be; it is elegant, and the result itself is nonintuitive. The result implies that if we let $P(n)$ denote the number of primes less than or equal to n, then $P(n)$ tends to infinity as n tends to infinity. [$P(2) = 1$ because there is exactly one prime less than or equal to 2, namely, 2 itself. $P(12) = 5$ because there are exactly five primes less than or equal to 12, namely, 2, 3, 5, 7, and 11.] The deep prime number theorem of Hadamard and De La Vallee Poussin gives the *rate* at which the function $P(n)$ tends to infinity. Euclid told us in 300 B.C. that $P(n)$ *tends to infinity*. Hadamard and De La Vallee Poussin told us the *rapidity* with which this happens. But between Euclid and the prime number theorem there was a lapse of two millenia. Euclid proved with an elementary argument that the prime numbers march out to infinity as the counting numbers 1, 2, 3, . . . march out one by one. Hadamard and De La Vallee Poussin had to reach down deep into the theory of complex analysis to establish the *pace* of the march of the prime numbers. What they told us is that $P(n)$ "grows" like $n/\log(n)$. [The symbol $\log(n)$ denotes the "natural logarithm of *n*." Think of log(*n*) as a value which grows as *n* grows, although not nearly so fast.] This remarkable result provides not only the answer to the question of the growth of the prime numbers but a connection between the discrete mathematics of the integers and the continuous mathematics involved with the logarithm. The method of proof solidified forever the connection between two seemingly unrelated fields of mathematics: number theory and complex analysis.

Complex analysis routinely provides connections between seemingly unrelated areas of mathematics or unrelated quantities. One such relation which is far less deep than anything connected with the prime number theorem follows from Euler's identity of 1748. Euler proved that for all real θ (theta)

$$e^{i\theta} = \cos\theta + i \sin\theta.$$

(The eighth letter of the Greek alphabet, θ is traditionally used in connection with trigonometric functions.) This equation expresses a relation between *functions*—the complex-valued exponential function on the left and the ordinary real-valued sine and cosine functions of trigonometry. (A function is simply a rule f which associates with one number, say x, another number denoted by $y = f(x)$. We have already looked at polynomial functions like $p(x) = x^2 - 2$. Trigonometry is largely the study of the two "rules" called "sine" and "cosine" which associate with a real number θ another real number y indicated by $y = \sin\theta$ or $y = \cos\theta$. These particular rules—as you may remember—are usually given by the ratios of lengths of sides of right triangles.) If we now take $\theta = \pi$ and remember that $\sin\pi = 0$ and $\cos\pi = -1$, we get

$$e^{i\pi} = -1. \tag{I}$$

This last result is interesting in itself and you can find it in several places in the mathematical literature (for example, it appears in the book *The Problems of Mathematics* by Ian Stewart, who points out the implication $\log(-1) = i\pi$). However, almost any mathematician who sees equation (I) will feel an irresistible impulse to add 1 to both sides. And, when he yields to this impulse, he obtains

$$e^{i\pi} + 1 = 0. \tag{B}$$

Equation (I) and equation (B) are exactly equivalent in that they contain the same information. That is, equation (I) is valid if and only if equation (B) is valid (and we know that equation (I) holds because Euler proved it). But the two equations live in different parts of the aesthetic world. Equation (I) is merely interesting; equation (B) presents a mathematical expression of great beauty. Notice.

The five most important constants in mathematics are the numbers e, i, π, 1, and 0. (There is no doubt of this; just stop any 100 mathematicians and ask them.) Moreover, the most vital relation in mathematics is the relation of "equality" and the paramount operations are addition, multiplication, and the operation called "exponentiation." Equation (B), as you can see, contains all of these things and *nothing else.* The equation portrays *completeness* because it contains these important mathematical concepts and, moreover, it contains nothing extraneous. (Thus, equation (B) satisfies something I call the *aesthetic principle of minimal completeness.*) Let us now notice the difference in the aesthetic values of equations (I) and (B) and understand that mathematicians are motivated more by the *beauty* of the latter equation than by the *information* of the former.

TWO PROBLEMS

Number theory is the branch of pure mathematics which deals with problems involving the integers. Such problems are relatively easy to state and, because of this, mathematical work in number theory appears from time to time in the public press. An example, recently reported, was an alleged proof of a 350-year-old conjecture known inaccurately as Fermat's Last Theorem.

Pierre de Fermat (1601–1665) studied law in Toulouse and worked most of his life as a civil servant. Consequently, descriptions of his mathematical researches usually refer to him as an "amateur mathematician." However, there was nothing amateurish about his mathematics and he contributed to many fields including coordinate geometry and probability. His work concerning maxima and minima of real-valued functions formed a sort of pre-Newtonian calculus. (A maximum of a function is just the largest value the function attains.) But he is best known now for his research in number theory, a field which Fermat founded single-handedly. Much of Fermat's fame springs from the simple fact that there exist infinitely many integers x, y, and z satisfying the equation

$$x^2 + y^2 = z^2. \tag{P}$$

Such integers are called "Pythagorean triples" because of their relationship to right triangles. An example of a Pythagorean triple is (3, 4, 5) because $3^2 + 4^2 = 5^2$. The exis-

tence of a single such triple implies the existence of infinitely many since, if x, y, and z satisfy equation (P), so do
the three integers ax, ay, and az, where a is *any* integer.
(Just multiply both sides of (P) by the number a^2.) Fermat
looked at the more general equation

$$x^n + y^n = z^n \qquad \text{(F)}$$

and asserted that there exist *no* integers x, y, and z satisfying equation (F) whenever n is an integer greater than 2.
Fermat made this assertion in a note written in the margin
of a book and claimed he had found a "remarkable" proof
but, unfortunately, the margin was too small to contain it.
More than three centuries have now passed but no proof
has been found, nor has a counterexample been produced
to show that Fermat's statement is incorrect. Many great
mathematicians have attempted to produce a proof—so
many that it has been said: "The history of mathematics
would have been different had Fermat's book wider
margins."

Perhaps. But I know of no mathematician who believes Fermat actually had a proof to his conjecture. Probably he simply erred and believed he had a proof when he
did not. Moreover, few mathematicians—at least until recently—believed the Fermat conjecture to be worthwhile.
The great Gauss himself worked on the problem in the eighteenth century and showed Fermat was correct for $n = 4$.
But, when the Paris Academy offered a prize for a complete
proof, Gauss shrugged it away saying the problem held no
value for the rest of mathematics. In the nineteenth century, the German mathematician Ernst Kummer described
the problem as more of a "curiosity" than a "pinnacle."

But work on the problem continues. In 1983, Gerd
Faltings proved that if there are solutions to equation (F),

then there are only finitely many solutions for each exponent n. So, if Fermat is wrong then he is only finitely wrong. Samuel Wagstaff and Jonathan Tanner, using a computer, showed Fermat to be correct for all values of n up to 150,000.

These are partial results and the general problem remains open. And it still has allure for most mathematicians. Just recently, an announcement appeared in the popular press that the conjecture had been solved and articles on the "solution" jumped out at you from journals like the *New York Times, The Chronicle of Higher Education,* and *Time* magazine. The celebration may have been, nonetheless, premature. As I write this, "fundamental errors" have been found in the proposed proof and Fermat's conjecture remains open—the prototype of a mathematical problem that is easy to state and to understand but, so far, impossible to solve. Mathematicians now have sound reasons to believe Fermat's Last Theorem true. But, unless Fermat had actually proved it, no human yet has.

A second problem, now solved, which routinely attracts the attention of expositors of mathematics because of its apparent simplicity, is The Four Color Problem. This problem—now more correctly known as the Four Color *Theorem* since it has been proved affirmatively—is:

> Can any map drawn in a plane be colored with four or fewer colors so that no two countries having a common boundary curve are of the same color?

A graduate student, Francis Guthrie, proposed this problem in a letter to his brother in 1852. Evidently, Guthrie received no help from his sibling and he presented the problem to his distinguished University College, London, professor, Augustus de Morgan; de Morgan also failed to solve the problem and passed it along to William Rowan Hamilton, who could not solve it either. And so it went from mathematician to mathematician for more than a century with no proof found that four colors suffice and no map produced which required more than four colors. Along the way, many mistakes were made and many first-class and nearly first-class mathematicians announced solutions which later were shown to be false. Ian Stewart describes the problem as "notorious" and claims that, for the 124 years it remained open, it was the most easily stated but most difficult question in mathematics.

The problem turned into a theorem in 1976 when Kenneth Appel and Wolfgang Haken of the University of Illinois showed—with the help of a computer—that the answer to Guthrie's question is "yes." (In celebration, the University of Illinois at Urbana adopted a postmark that canceled postage with the phrase: "Four Colors Suffice.") The Appel–Haken proof caused, however, considerable controversy.

A gross oversimplification of the Appel–Haken proof, but one which provides insight to their method and to the reasons for the controversy, goes like this: to prove that there is an affirmative answer to the four color question you must prove that *any* allowable map can be colored with four or fewer colors. Simply to show that the vast majority of maps can be so colored might be psychologically satisfying but would provide nothing in the nature of a *proof.* Proof requires generality. You must deal with *all* maps.

Nor would it be sufficient to demonstrate that all maps belonging to a certain class of maps could be colored with four colors. Unless, of course, you first proved the result: *Any map can be colored with four colors provided that all maps in the given class can be so colored.* Then, when you have settled affirmatively the question for this class of maps you have settled it for all maps. This, in essence, represents the method used by Appel and Haken. Using prior work of other mathematicians such as Kempke, Birkhoff, and Heesch, they reduced the general problem of coloring *any* map with four colors to the particular problem of coloring a finite set of maps belonging to a certain class. The problem then became one of checking, one-by-one, each map in this particular class. However, the number of maps that had to be cleared, while finite, was enormous. Checking them all lay beyond the range of human computation. So, Appel and Haken turned to Illinois's IBM 360 computer and had it check the special cases. After months of real time and a thousand hours of computer time the machine gave the answer: yes.

The deed was done. Or was it? Appel and Haken had produced a new kind of "proof." A historically significant and deep mathematical result has been proved in a manner which cannot be reviewed by mathematicians. The computer's role stands, in its entirety, outside of human evaluation. The computer did too much, made too many calculations too fast, for its work ever to be examined by mere mortals. If the computer erred, we are likely never to know (unless, of course, someone produces a map which requires five colors).

The Appel–Haken method attracted the attention of philosophers. Thomas Tymoczko, writing in the *Journal of Philosophy,* suggested that acceptance of this method requires fundamental changes in our concept of "proof." He wrote[13]:

> A proof is a construction that can be looked over, reviewed, verified by a rational agent. We often say that a proof must be perspicuous or capable of being checked by hand. It is an exhibition, a derivation of the conclusion, and it needs nothing outside itself to be convincing. The mathematician *surveys* the proof in its entirety and thereby comes to *know* the conclusion.

The Appel–Haken demonstration cannot be "looked over" and, consequently, cannot be known. Tymoczko considers it *not* a proof.

On the other hand, the philosopher Israel Krakowski quickly took issue with Tymoczko and argued that, though the Appel–Haken method is "distinctive," it "does not raise any new issues of philosophical importance." Tymoczko says the proof is not surveyable. Krakowski insists the computer has surveyed it in "step-by-step fashion." And, says Krakowski, to suggest the computer has done otherwise is "chauvinism" (surely the first documented step toward assigning victim status to the country's collection of digital computers).

Tymoczko says the Appel and Haken solution "introduces experimental methods into mathematics." Krakowski replies that mathematics is already experimental or, at least, empirical. Ian Stewart implies that mathematicians are either indifferent or else feel the method of solution shows "it wasn't a good problem after all."

Probably nothing is amiss and the proof is correct. Stewart believes it true and says he knows of "no mathematician who currently doubts the correctness of the proof." Moreover, asserts Stewart: "It is also much less likely that a computer will make an error than a human will."[14] Maybe. But we will never know. For the question of "surveyability" of the work of computers holds interest only for the philosophers. The rest of us know that complicated computer output goes unread, let alone *surveyed.* Computer printouts longer than five pages, we know, go directly from printer to desktop to file cabinet.

Appel and Haken may have given us a glimpse of the future. A future in which deep theorems routinely will rely for their proofs on the checking of millions of special cases by high-speed, unmonitorable computers. In the book *Mathematics Today,*[15] Appel and Haken indicate this might come about. And I gather it is a future they welcome.

But I do not. For it is a future without elegance, a world

of disfigured mathematics. Truth may choose to live in
that world but beauty will not. There, the *art* of mathematics will, like Prospero's spirits, vanish into thin air. In this
brave new world the magician will snap his fingers and
mathematicians will be changed from poets into
cabinetmakers.

Applied Mathematics

The game of pure mathematics is played in your mind. You keep track of the game's moves by means of symbols you write on paper. As the game progresses and abstraction piles upon abstraction, symbols come to represent collections of other symbols. Metaphors replicate themselves and what began as a study of analogies between objects finally develops into the manipulation of analogies of analogies. At this stage the mathematics takes a life of its own and itself creates new objects of thought. And, while the beginning axioms and definitions may have been set down as a reflection of reality, you arc now far from them. You are in the open sea of ideas. The horizon of reality vanished behind you days ago. Surely, these abstruse notions can have no worth beyond their value toward the creation of yet more mathematics. They can be of no use in the real world.

But they are. Often. Alfred North Whitehead wrote: "The paradox is now fully established that the utmost abstractions are the true weapons with which to control our thought of concrete fact."[1] And—in a dictum used for two and one-half centuries to justify the pursuit of pure mathematics—Galileo said: "The great book of nature can be read only by those who know the language in which it was written. And this language is mathematics."

The intellectual realm characterized by the use of mathematics in the handling of concrete facts and the understanding of nature is called *applied mathematics.*

Obviously, the scope of applied mathematics is broad and the boundaries between this discipline and others such as physics or economics will be necessarily vague. Whether, for example, the person who writes a research paper in mathematical physics describes himself as a physicist or an applied mathematician depends mainly on what is his main line of work. If he always works in physics then he is a physicist. If his main interest lies in the mathematics and in its application to physics, and quite possibly other fields, then he calls himself an applied mathematician.

Here, we need not draw such fine distinctions. For our purposes, it will suffice to think of applied mathematics as the *process* by which mathematics is connected to the real world so as to produce new information about Whitehead's "concrete facts." We will see that this process falls naturally into three pieces: the *applicability* of certain kinds of mathematics, the *mathematics* itself, and the *application* of the mathematics. ("Applicability" roughly refers to the "usefulness" of certain kinds of mathematics while "application" refers to the "process" of using the mathematics.) Moreover, the act of doing applied mathematics involves mathematics only in the middle part. At each end, the process of applied mathematics falls outside the limits of pure reason. Eugene P. Wigner, a renowned physicist and Nobel laureate, described these end procedures as "unreasonably effective" and said: ". . . the enormous usefulness of mathematics in the natural sciences is something bordering on the mysterious and there is no rational explanation for it."[2]

The process of applied mathematics can best be described through examples and by means of diagrams which show exactly where the three parts of the process separate. In this way we can distinguish clearly when applied mathematics is mathematics and when it becomes witchcraft.

Let's consider first a simple and well-known example: a struck golf ball flies off the ground at a given angle with a known initial velocity. How far does it go?

The first step in the analysis requires the identification of the relevant factors. Evidently, neither the golf club that struck the ball nor the person who swung it are significant since we are given both the velocity and the angle with which the ball leaves the ground. Next, we inquire whether the size, shape, or weight of the ball matters. Clearly, the size and shape of the ball must be considered if air resistance to its flight is to be a factor. But if we assume the ball has aerodynamically effective dimensions—as we expect of an actual golf ball—the size and shape can be ignored.

Since we have chosen to ignore air resistance, we *assume* that the only relevant force that will act on the flying ball will be the force of gravity. Consequently, we expect the ball's weight to be a relevant factor since the force with which gravity attracts an object is the weight of the object. And, more fundamentally, the weight of an object equals the product of its mass m and the acceleration caused by the force of gravity g. So, we must know the weight w of the golf ball—or equivalently its mass m—in order to proceed. Let's suppose we know the mass, m, of the ball.

Next, we decide which of the "laws of nature" govern the flight of the ball. In this case we need only the ordinary laws of motion and, in fact, we need only Newton's *second law* which tells us that the force on the ball will equal the product of its mass and its acceleration. If we let v denote the unknown velocity of the ball at an arbitrary moment of its flight and if we remember that acceleration is the calculus "derivative" of velocity with respect to time t, the second law allows us to write a differential equation which reduces to:

$$\frac{dv}{dt} = g. \qquad\qquad (D)$$

(Think of velocity as the speed at which you drive and acceleration as the rapidity with which you change speeds.)

We need not concern ourselves with the technical meaning of equation (D) nor with the form of its solution. (The equation expresses, in precise mathematical terms, the statement: "The rate of change of velocity with respect to time equals the gravitational acceleration g." It is called a "differential equation" because dv/dt is a calculus concept called "the derivative of v with respect to t.") But we should make three observations:

First, equation (D) contains no m. The mass has dropped out of the calculation. Our expectation that the weight of the ball would be relevant to the analysis was incorrect. The assumption that the ball would be affected *only* by the force of gravity made weight, as well as size and shape, irrelevant. Second, equation (D) is not quite as simple as it appears. The golf ball is moving along a curved path in a "plane of flight" and not moving simply on a straight line. Consequently, the position of the ball at any moment is given by two coordinates, x and y, as shown in Figure 4. Equation (D) is a *vector* equation and may be replaced by a pair of differential equations involving the ball's (x, y) position coordinates. (Think of a vector of an

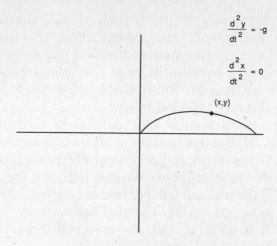

FIGURE 4. The flight of a golf ball.

arrow with its notched end fixed at the origin and its point on the moving ball. The arrow then stretches and rotates as the ball moves. Equation (D) describes this movement.)

Finally, we must observe that the problem of examining the flight of the ball has been reduced to an analysis of equation (D). By solving this equation we can find the answer to the original question concerning the distance traveled by the ball. Equation (D) is called a *mathematical model.*

We produced the mathematical model for the golf ball's flight by using the relevant factors and the appropriate laws of nature to write a differential equation. While the original problem concerned a real-world situation—the flight of a real golf ball—the mathematical model lives outside the world of reality. The model—equation (D)—is a mathematical object, an abstraction, and it lives in the world of mathematics. In order to solve equation (D), we manipulate the symbols according to the laws of logic and the rules that apply to the branch of pure mathematics known as "ordinary differential equations." No part of the process of solving the equation involves *any* aspect of reality. To solve the equation is to do *pure mathematics.*

When we have the solution at hand we use it to determine the distance traveled by our abstract golf ball. Fine. We have a mathematical ball that moves according to a mathematical equation and the solution tells us how far the ball goes. No problem. But now we apply the solution to the model to the real-world ball. We conclude that the distance traveled by the real ball will be the distance *predicted* by the solution to equation (D). The real ball—we fervently believe—will behave as the abstract ball behaves. And, wonder of wonders, it does.

The mathematics began when we wrote the equation which led to equation (D). The mathematics ends when equation (D) is solved. The initial step of forming the mathematical model by identifying the relevant factors and fixing on the appropriate physical laws requires knowledge and experience and can be described as the "art of modeling." The final step of associating the mathematical truth of the model with the real-world golf ball lies outside mathematics and maybe outside reason. What is remarkable is that it works. Unreasonably, irrationally, uncannily—it works.

And it works everywhere. Take the space shuttle to the back side of the moon. Wobble out in your spacesuit and hit a golf ball across the lunar landscape. Tell me its initial velocity and tell me the moon's gravitational constant. Then I, sitting in my study in Pennsylvania, will write mathematical symbols on yellow paper and manipulate them around as if I owned them. When I'm done, I'll *compute* how far the ball goes.

You will have lost the ball in the darkness of the

moon's back side. But don't bother with flashlights. Just call down. I'll tell you exactly where to find it.

Now let's describe the applied mathematics process through a diagram in order to see clearly the separate parts. Figure 5 shows the mathematical world of Figure 1, with the "countries" removed, and, next to it, another rectangle representing the real world. The aim of science is to understand the rectangle on the left—the real world. Whitehead's "concrete facts" and Galileo's "book of nature" refer to phenomena that live here.

Let's consider a piece of the real world which we wish to understand. This "piece" is represented by the shaded region in the real-world rectangle of Figure 5. You may think of this "piece" as a problem of the type we've already examined. Perhaps we want to understand the flight of a golf ball or some, more general, projectile. Or, perhaps, we want to comprehend the spread of a certain epidemic. Maybe we are concerned with the forces that bind together the particles that make up a carbon atom. It doesn't matter. The shaded region denotes some real world phenomena we want to understand in a scientific manner. We want to de-

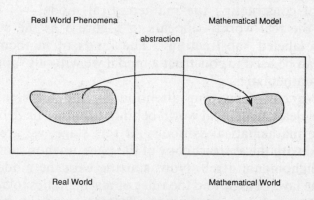

FIGURE 5. Model making.

velop a theory regarding the phenomena which will agree with what we have observed about it in the past and will, if we do it right, predict future behavior.

To "control our thought" about the phenomena we need to first develop a symbolic metaphor for it. This first step is the construction of a mathematical model for the piece of the real world under investigation. This step requires that we construct, in the mathematical world, an abstract "copy" of the real-world phenomena. In the case of the golf ball, the mathematical model consisted of a single differential equation. Usually, the mathematical model is far more complicated and may involve systems of equations and inequalities together with a set of "boundary conditions." (Boundary conditions usually describe the "initial state" of the real-world phenomena. For the golf ball, the boundary conditions are the velocity and angle with which the ball left the ground.) But, as in the case of the golf ball, the mathematical model is constructed only after relevant factors have been isolated and the appropriate governing physical laws identified. The mathematical model for the given problem is indicated by the shaded region shown in the mathematical world rectangle of Figure 5. The process of constructing the mathematical model—of going from the real world—appears in Figure 5 as the curved arrow labeled "abstraction." The kinds of mathematics that can be used to construct a model we will call "applicable" mathematics.

We now focus our attention on the mathematical model, leaving the real world behind. What we have before us are mathematical symbols. At this stage, we are concerned with the abstract copy of the phenomenon, not with the phenomenon itself. Now, starting with the model, we use the laws of logic and the rules of mathematics to *deduce* other properties it may possess. That is, we manipulate the mathematical symbols to discover previously unknown "facts" about the model. In a particular case, this process will involve the solving of equations or the summing of certain infinite series or, perhaps, the evaluation of complex integrals. But, at each stage of the deduction, our arguments can be phrased in the form of "implications," that is, in logical statements of the type "*p* implies *q*." This is the *analytic* stage of the process and it is the stage of *pure mathematics.*

When the analysis is finished and we have pushed the mathematics as far as we can, we will have discovered "truths" about the model which we did not know at the outset. In the case of the golf ball, for example, the analysis of the model required the solution of a differential equation. And, from the solution, we learned how far the ball would travel—a "fact" we did not know when we first set down the model. In the diagram, we indicate this new knowledge of the model by a "fattening" of the region in the mathematical world of Figure 6. This enlargement appears in Figure 6 as the diagonally shaded ring. The curved arrow labelled "analysis" denotes the pure mathematics manipulation which results in the new information about the model.

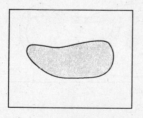 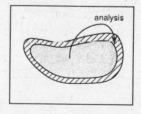

 Real World   Mathematical World

FIGURE 6. Manipulating the model.

It must be made clear that these new truths are mathematical truths. What the pure mathematics process gives us is new information about the mathematical model. The analysis tells us "facts" about the model we did not previously know. In the case of the golf ball, the analysis told us, for example, how far it travels. But we must remember that it is the "mathematical ball," not the real ball, to which the analysis applies. The new information, like the model itself, lives in the mathematical world, not the real world, of Figure 6.

The mathematics ends here. The next step—that of "applying" the new mathematical truths to the real-world phenomena—is not mathematics. Mathematics requires rules of inference. Yet, there are no rules which allow you to infer that the mathematical truths carry over the real-world phenomena. This step requires magic and the true belief that it will work. E. P. Wigner calls it an "article of faith."

The completed diagram, Figure 7, shows the entire loop including the final "application" step which we believe provides new "facts" about the real-world phenomena under consideration. This new information appears in

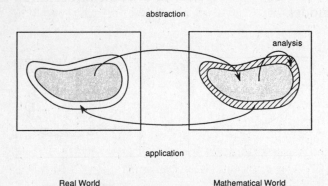

FIGURE 7. The applied mathematics process.

Figure 7 as the enlarged real-world region. In the case of the golf ball, the enlarged real-world region includes such new "facts" as how far the ball goes, how high it goes, and its exact position at any instant of time.

ERRORS

Putting aside for the moment the "leap of faith" required in the last step of the applied mathematics process, we see two places in which mistakes can occur. The first possibility of error occurs in the construction of the mathematical model and the second possibility for error appears in the mathematical analysis.

The construction of the model requires that the relevant factors are properly identified and accounted for. It is easy to go wrong here. For example, we chose to ignore air resistance in our analysis of the golf ball's flight on the grounds that the ball would be of such aerodynamical design that the resistance would be negligible. This may have been a mistake. Perhaps we have a misshapen ball or perhaps it is flying into the teeth of a sharp wind. If so, there is even less reason for us to believe the output of the applied mathematics process will provide useful information about the flight of the ball.

Mathematical models often require refinement and tuning. You may begin with a simplified model, as we did with the ball, and find it to be inadequate by noticing that the predicted "truths" do not agree with the observed behavior of the real-world phenomena. Real golf balls, for example, may not fly as the model predicts. In that case, the model is refined—usually through reconsideration of the relevant factors—and the process is repeated.

The refinement of the model may make the mathematical analysis more difficult. To add air resistance to the golf ball model means the differential equation (D) must be replaced by one more complicated and, hence, more difficult to solve. In fact, a complete consideration of all possible forces affecting the ball may lead to a differential equation incapable of solution in closed form. In this case, one must turn to approximate, numerical solutions which usually require the use of high-speed computers and, invariably, introduce further error. The *art* of modeling consists of the technique, learned through experience, which allows a realistically complete consideration of relevant factors while, simultaneously, yielding a model amenable to mathematical analysis.

But no matter how refined our art or how deep our experience, the model will be at best a representation of the real-world phenomena and never an exact copy. (Any real-world phenomena will be too complicated for exact modeling. We model the factors that are relevant and the factors we can manage. In the golf example, our model ignores such factors as the effect of air currents on the ball's flight.) Because of this we can never expect the output of the applied mathematics process to give exact results about the real world—no matter how fervently we believe in Wigner's article of faith. Approximation to reality is the best we can do. Error is inevitable.

We must, consequently, exercise extreme care to ensure that further error does not occur in the mathematical analysis itself. We must avoid ordinary mathematical mistakes in the analysis. But we must go further. The analysis must be carried out with as much rigor and precision as we can manage. We must not compound the error, already inherent in the process, by allowing additional, subtle error to creep in through cracks caused by casual handling of the mathematics.

It is on this point—the degree of rigor required in the

mathematical analysis—that pure mathematicians and applied mathematicians disagree most sharply. Pure mathematicians stand on the side of precision and rigor for that is what mathematics is about. And, in the analysis stage of the applied mathematics process, what occurs *is* pure mathematics. The applied mathematicians, on the other hand, are interested in the end result of the process, the *application,* and are inclined to treat the mathematics in whatever manner—rigorously or offhandedly—that gets them to it most quickly.

Consider, for example, a simple pendulum consisting of a small weight hanging at the end of a light string. The weight is pulled slightly to one side and released. How does the pendulum move thereafter?

A standard mathematical model for this phenomenon consists of the single second-order differential equation

$$\frac{d^2\theta}{dt^2} = \frac{-g}{L} \sin\theta. \qquad (R)$$

Here, L represents the length of the string, g the acceleration due to gravity, and θ denotes the angle of deflection of the pendulum at an arbitrary time t (see Figure 8). We are not concerned with the development or the exact meaning

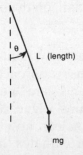

FIGURE 8. A simple pendulum.

of equation (R). But you should notice that, as in the case of
the golf ball, the mass m of the weight at the end of the
string does not appear in the equation. Consequently, the
mass does not represent a relevant factor in this mathematical model.

The presence of the trigonometric term $\sin\theta$ makes
equation (R) difficult to solve. But, for small values of θ,
$\sin\theta$ and θ are nearly equal. And the problem becomes
greatly simplified if you replace $\sin\theta$ by θ so that equation
(R) reduces to:

$$\frac{d^2\theta}{dt^2} = \frac{-g}{L}\,\theta. \qquad (S)$$

Equation (S) may be easily solved by standard methods and its solution provides an explicit relationship between θ and t. Thus, the solution shows how the angle θ
varies with time and, allegedly, describes the motion of the
pendulum—at least for small values of θ.

An applied mathematician will make this simplifying
replacement without hesitation. Moreover, when he has
finished, he will often refer to the solution of equation (S)
as "the equation of motion of the pendulum," as if no other
solution or procedure were possible. He speaks of "the
equation" as if he has found "the truth."

A pure mathematician, on the other hand, will be troubled, for several reasons:

- Although the quantity $(\sin\theta)/\theta$ truly (in a precise
  mathematical sense) approaches 1 as θ tends to 0, the
  notion that $\sin\theta$ and θ are "nearly equal" for "small"
  values of θ remains vague. Without more rigorous
  analysis, it is unclear how closely the simplified
  equation (S) approximates the real equation (R).

- Moreover, even if the approximation of equation (R) by equation (S) can be made precise, you cannot conclude *a priori* that the *solution* of equation (S) approximates the *unknown solution* of equation (R). Resolution of this point might be as difficult as dealing directly with equation (R). And failure to resolve it means that we must simply *believe* the solution of (S) will somehow be close to the solution of (R), thus introducing a second "article of faith" into the process.

But, beyond these specific criticisms, the pure mathematician frets over the *casualness* of the analysis. Once the model has been formed, you are in the analysis stage of the applied mathematics process (see Figure 7). And what happens here—or should happen—is pure mathematics. The building of the model requires intuition and experience and, inevitably, approximation as you construct a copy of real-world phenomena over in the world of ideas. And, at the end of the loop, you are forced to accept Wigner's article of faith in order to believe in the efficacy of the applied mathematics process. But in the middle there are *rules* because what you do in the middle is pure mathematics. And pure mathematics means precision if it means anything at all. *Imprecision* is contrary to its very nature.

The applied mathematician absolves himself by pointing out that the simplification "works" in the sense that the solution of equation (S) agrees—at least for small initial deflections—with the *observable* behavior of real-world pendulums. And, he claims, it is this matching of theory with observation that gives the model scientific status and not the level of mathematical precision that produced the theory.

Perhaps. But the pure mathematician knows that a

more precise analysis of equation (R) would also yield an observably correct theory; correct, probably, for a wider range of values of θ than the simplified theory. Moreover, the pure mathematician worries over the introduction of an "end justifies the means" principle into mathematics. He knows, for example, that you can divide 64 by 16 by writing the appropriate fraction and simply canceling the 6 in the numerator and denominator:

$$\frac{\cancel{6}4}{1\cancel{6}} = \frac{4}{1} = 4.$$

But he also knows that you get into deep trouble with this method if you try it, for example, on 64 divided by 26.

The applied mathematician emphasizes the application; the pure mathematician reveres the analysis. The applied mathematician wants an answer. He has little patience with mathematical subtleties. The pure mathematician wants precision. Imprecision provokes him the way a rabbit provokes a tied beagle. They are like the two men in the Escher print who tread the same step on the same staircase in the same direction with one going up while the other goes down. Contact between them seems impossible. They live in separate worlds.

SEPARATION

The applied mathematics process outlined in Figure 8 involves pure mathematics in the middle stage but it may not involve a pure mathematician. In fact, in the great majority of cases no pure mathematician participates. Ordinarily, the process goes forward under the hand of an applied mathematician or a physicist, or other scientist, or, perhaps, a team of such individuals. If the problem under consideration is significant—say the design of a control system that will land the first manned space shuttle on Mars—the process will take place repeatedly and over a long period of time. And, at the finish of each completed loop of Figure 7, the output of the application stage—"the theory"—will be tested either directly through observation or indirectly through such methods as computer simulation. If the output fails the test, the model may be refined and the process repeated until an acceptable theory emerges. The model's accuracy depends on the rigor of the analysis. However, the level of rigor at which the analysis stage is conducted is likely to depend, not on what the mathematics *requires,* but on the predilections of the people who conduct it. Since they are not likely to be pure mathematicians, the danger exists that the analysis will be too casual to ensure the development of an acceptable theory without an inordinate number or repetitions of the complete process.

It would seem desirable to increase the participation of pure mathematicians in applied mathematics processes. After all, pure mathematics constitutes a significant part of the process so it is natural to expect the people who know it best to participate in the process. Moreover, the involvement of pure mathematicians would undoubtedly raise the precision level of the analysis and quite possibly make the applied mathematics process more effective, thereby reducing the need for expensive and time-consuming repetitions of experimentation or computer simulation.

At this moment, there exists a growing collection of reports and articles indicating a deepening involvement of pure mathematicians in the applied mathematics process and a blurring of the boundaries between these two areas of mathematics. Consider, for example, the 1986 report of the

Panel on Mathematical Sciences, initiated by the National Research Council. The panel concluded: "Mathematics is unifying internally. The division between pure and applied mathematics that developed in the first part of the twentieth century and allowed for rapid development of new fields is now disappearing."[3] Also, a recent conference held at the National Academy of Sciences called, "Mathematics: The Unifying Thread in Science" had as its theme the interfacing of pure mathematics and the applied mathematics found in the physical sciences. Moreover, a current article by a past president of the Mathematical Association of America emphasizes the "one system" nature of the whole spectrum of mathematical sciences and points particularly to the "unity and applicability of mathematics research" in areas including computational statistics, mathematical biology, and nonlinear dynamics.

I do not wish to take issue with these optimistic views about the increasing unification of mathematics and of mathematicians. If it happens, both pure and applied mathematicians will be strengthened, mathematics education will be improved, and society will benefit. Unification has my best wishes. The report distributed at the above-cited conference says[4]:

> The unification that is taking place within mathematics is obvious to people in the field. . . . The symbiotic relationship between mathematics and its areas of application is ever deepening as more areas of science and engineering become almost indistinguishable from subareas of mathematics . . .

Perhaps. But my own experience shows little "unification" of the people in the various areas of mathematics and, in particular, none whatever between those mathematicians who adamantly place themselves on opposite sides of the pure mathematics–applied mathematics boundary. Almost without exception, I find pure mathe-

maticians and applied mathematicians to have vastly differing attitudes and expectations. The most they have in
common is a lack of appreciation for each other's work.
Pure mathematicians live to *create* mathematics; applied mathematicians exist to *use* mathematics. Much of
the mathematics used in the application process is well-
known and routine. This kind of mathematics can be called
applicable mathematics. It is what the world calls "useful"
mathematics. The opportunity to work with existing applicable mathematics excites a pure mathematician exactly as
does an invitation to watch paint dry.

APPLICABLE MATHEMATICS

The analysis stage of the applied mathematics process
involves the study and manipulation of a mathematical
model in order to obtain "facts" about the model which
were not known at the outset. If this step routinely—or
even frequently—required the creation of new mathematics, you would have no trouble whetting the interest of pure
mathematicians. But, alas, it invariably does not.

Still, applicable mathematics—the mathematics used
to build the model—does not have to be old. New mathematics may be required. Moreover, each piece of pure
mathematics—no matter how abstract—is theoretically *applicable*. And these two points are exactly what is emphasized in reports such as the previously cited document
"Mathematics, the Unifying Thread in Science." The first
point implores pure mathematicians to become involved in
the applied mathematics process and the second justifies
the work of those who do not. Let's look at these points one
at a time.

The quintessential example of new mathematics arising from an attempt to model real-world phenomena is the creation of calculus by Isaac Newton. Mr. Newton, in about 1660, considered the dynamics of Galileo which concerned itself with the straight line motion of bodies under constant acceleration. Newton wanted to study the more general motion which results when the acceleration is allowed to vary. To model this phenomena, he found he needed a new kind of mathematics which would describe precisely and simply the concept of rate-of-change of position with respect to time. In fact, what he most needed was mathematics to deal with rate-of-change of rate-of-change, for in this notion lies the essence of the concept of variable acceleration. (Velocity is rate-of-change of position. Newton wanted to understand *acceleration*, which is the rate-of-change of velocity.) The result was Newton's great *Principia* containing his ideas on gravitation. The famous by-product was that part of mathematics now called *calculus*.

This story of a real-world motivation leading to the creation of a vital branch of pure mathematics is justly famous. But it is too singular to stand as an example for modern mathematicians to emulate. Newton remains the greatest genius of science and the *Principia* stands as the paramount scientific publication. The probability that a contemporary pure mathematician will—by pursuing *any* scientific problem whatever—be led to the discovery of mathematics of comparable value is exactly zero. (This does not mean that it cannot happen but rather that it is *extremely* unlikely.)

Incidentally, the calculus was also created, independently and almost simultaneously, by Gottfried Wilhelm Leibniz. Leibniz—whose symbolic notation was far superior to Newton's and is still used today—seems to have been motivated more by philosophical and geometric questions than by real-world physical problems.

One can find modern instances where the investigation of real-world phenomena has yielded new mathematics at the analysis stage of Figure 7. But in the bulging net called applied mathematics the examples are as rare as freshwater pearls. To call attention to them is to press a point. And to imply, as does the "Unifying Thread" report, that because "many of the mathematical constructions that mathematicians have developed over the centuries were originally inspired by scientific problems,"[5] we can then expect significant carryover of this kind of productive inspiration to modern-day pure mathematicians results only in a loss of credibility.

A stronger case can be made for the second point— that any piece of pure mathematics may ultimately turn out to be applicable mathematics. Look again at Figure 7. *Applicable mathematics* is defined to be the mathematics resulting from the creation of the mathematical model of the applied mathematics process. We usually think of this process as moving clockwise with respect to time. That is, first we have the real-world problem, then we create the mathematical model, then we do the analysis on the model, and then we apply the result of the analysis back to the real-world phenomena. But the order could be changed.

For example, a pure mathematician might, for whatever reason, conduct a piece of mathematical research. This research leads to the creation of new mathematics and it may be indicated by diagram exactly as shown in the analysis stage of Figure 7. Now, we think of the diagonally shaded region in the mathematical world as being the piece of pure mathematics with which our researcher begins. The enlarged region indicates the new mathematical truths that have been discovered by the research and the curved "analysis arrow" denotes the research process.

This research—which may be completely abstract and

apparently unrelated to any physical phenomena—may later be found to be exactly the mathematics necessary to model a particular real-world problem. That is, the "abstraction arrow" of Figure 7 may sometimes be reversed; our pure mathematician's abstract research may later be found to have a *pre-image* in the real world.

This gives us a much more general notion of applicable mathematics. Namely, applicable mathematics is any piece of pure mathematics which has a pre-image in the real world under the abstraction arrow indicated in Figure 7.

If we think of the abstraction arrow as representing what the mathematicians call a "set function," i.e., a mapping of subsets of the real world into subsets of the mathematical world, then pre-image may be replaced by the more exact term "inverse image." In this sense, a piece of the mathematical world is applicable provided it has an inverse image in the real world under the function indicated by the abstraction arrow of the applied mathematics process.

Numerous examples of this more general kind of applicable mathematics exist. One famous and classical example involves the collection of plane curves known as conic sections. This family of curves includes all circles, ellipses, parabolas, and hyperbolas. The curves were first investigated in about 200 B.C. by a pure mathematician named Apollonius.

Apollonius of Perga was perhaps the last great mathematician of the golden age of Greek mathematics and his name is often linked with Archimedes and Euclid as representing the best of that period. Apollonius wrote extensively and one of his most celebrated works was a long treatise called "Conics." The work consisted of eight books and described systematically, beginning from first principles, most of the geometric properties of these curves. Apollonius defined the conic sections as the curves formed by the intersection of a plane with a double circular cone. When the plane is perpendicular to the main axis of the cone, the curve of intersection is a circle. As the inclination of the plane changes, the curve transforms from a circle into first an ellipse, and then a parabola, and finally a hyperbola having two branches (see Figure 9).

Apollonius showed that the conic sections could also be described analytically. In particular, he showed that— just as a circle can be considered as the path of a point moving in a plane at a fixed distance from a fixed point—

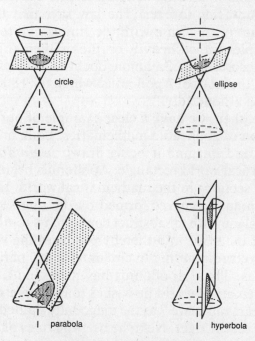

FIGURE 9. Conic sections.

an ellipse may be described as a point moving so that the sum of the distances from two fixed points is constant. The two fixed points are called the *foci* of the ellipse.

Apollonius's work was pure mathematics and his theorems on conic sections sat in the mathematical literature, in plain view but virtually unused, for almost 2000 years. In about 1600, Johannes Kepler revolutionized astronomy by formulating his famous three laws of planetary motion. Kepler's first law broke the celestial orbit tradition that had dominated astronomy since Eudoxus in 370 B.C. This tradition held that the basic celestial trajectory was a circle and that planetary motion was either circular or composed of combinations of circles. Kepler's first law set this supposition aside forever. The law says that the path of each planet is an *ellipse* with the sun as one focus. In his work, Kepler leaned heavily on the mathematics of Apollonius. According to Salomon Bochner, Kepler "utilized, crucially, the work of Apollonius, which for 1800 years had been lying around idly."[6]

Here, then, we have a clear example of our more general notion of applicable mathematics. Look once more at Figure 7 and imagine it being drawn *beginning* with the mathematical world rectangle. Apollonius begins by creating conic sections in the mathematical world. He does this by imagining the curves formed by the intersection of an abstract plane with an abstract cone. In his mind, he varies the tilt of the plane and thereby produces the various special cases corresponding to circles, ellipses, parabolas, and hyperbolas. Thus, Apollonius has identified a portion of the mathematical world he wishes to investigate. This portion appears in Figure 7 as the diagonally shaded area in the rectangle on the right. Next, he uses the rules of mathematics and the laws of logic to deduce other properties of these curves. For example, he determines the shortest distance from an arbitrary point to a given conic and discusses the existence of lines perpendicular to the conic. Most of what we know today regarding these curves was discovered by Apollonius, who worked, incidentally, without benefit of our modern symbolic mathematical notation. The truths established by Apollonius filled his eight books and are indicated in Figure 7 by the "analysis arrow" and the enlarged mathematical world region.

Centuries pass. The pure mathematics of Apollonius stagnates virtually unused, surviving only in Arabic and not yet translated into Latin. Along comes Kepler, who wants to study the real-world problem of planetary motion. This problem becomes the diagonally shaded area of the real-world rectangle in Figure 7. Kepler sees that he must build a mathematical model. And he finds the ancient pure mathematics of Apollonius to be exactly what he needs. Apollonius, Kepler saw, had "prefabricated" exactly the needed mathematics.

Consequently, the pure mathematics of Apollonius turns to "applicable mathematics" after a gestation period of almost twenty centuries. What Kepler found, with his model of the solar system, was the real-world pre-image of the pure mathematics of Apollonius.

To be sure, Kepler needed mathematics other than that of conic sections to complete his work. He, in fact, made use of infinitesimal methods and, thereby, contributed in some way to the development of the calculus. Kepler actually illustrates both of the points regarding applicable mathematics: the first, which says that the solution of real-world problems may stimulate the creation of new mathematics and the second, which states that any piece of pure mathematics is potentially applicable to some real-world problem.

But the importance of the contribution of Apollonius

cannot be overstated. Salomon Bochner says: "Kepler was a direct successor to Apollonius" and "without Kepler there would have been no Newton."[7]

Apollonius to Kepler to Newton; conics to planets to gravitation. A double-play combination leading from the pure mathematics of the golden age of ancient Greece directly to the great *Principia.*

Modern examples also exist of this second point regarding applicable mathematics. Einstein, for example, found that to describe his theory of general relativity he needed the curved-space geometry which had been developed decades earlier by the pure mathematicians Riemann, Gauss, Bolyai, and Lobachevsky. A second example is provided by the widespread use in modern physics of some highly abstract mathematics called group theory: a mathematical theory that deals with sets and a very general notion of multiplication of elements of these sets. This theory has been called "the supreme art of abstraction" and was developed by pure mathematicians—beginning with Galois in 1830—without consideration as to its possible utility. Its applicability was determined much later but the theory now is used routinely in the mathematical sciences. A third example comes from the theory of matrices, a piece of pure mathematics invented by James J. Sylvester and Arthur Cayley which allows the manipulation of huge arrays of numbers as if they were single elements. More than sixty years after the creation of matrices and the development of their theory, the physicist Werner Heisenberg used them in the mathematical modeling of what is now known as quantum mechanics.

None of these mathematical theories was developed for utilitarian reasons. Each was at the time of its creation an example of pure, pure mathematics. But each theory has become a practical tool used for understanding the real world.

Now we've come to something really spooky. The applied mathematics process *works*. Wigner's article of faith, which he and the rest of us require in order to make the final, mysterious step at the application stage of Figure 7, is invariably vindicated. Time and again, the applied mathematics process demonstrates the unreasonable effectiveness of mathematics in the natural sciences. Wigner's "miracle" repeats and repeats.

How can this be? The miracle itself is wonder enough. Wigner wrote: "The miracle of the appropriateness of the language of mathematics for the formulation of the laws of physics is a wonderful gift which we neither understand nor deserve."[8]

Yes. But if, as I claim, the motivation for the development of mathematics is primarily aesthetic, and not utilitarian, then the wonder compounds itself. We are talking about the *paradox of the utility of beauty*. And we are dealing with a miracle of second order of magnitude.

CHAPTER 5

Aesthetics

n the old days all of us studied philosophy because the twin concepts of liberal education and core curriculum were fixed in place and philosophy was part of what you should know if you were to be considered educated. And, in the beginning of Philosophy 101, we learned the four classical questions with which the subject deals:

> What is truth?
> What is reality?
> What is justice?
> What is beauty?

These four questions, our professor said, go by the appellations of *the cognitive, the metaphysical, the ethical,* and *the aesthetical.* What philosophy is all about, he told us, is to bring meaning to these great and fundamental issues and to establish progress and procedures toward their resolution.

I intend here to examine carefully the last of the four questions from the point of view of its intersection with mathematics. But, before I begin, I want to make two

points about the classical questions which were presented
to us as the core philosophy:

- Of the four, the question which seems to have received the least attention and remains in the most
  muddled state is the aesthetic question.
- Mathematicians do mathematics for aesthetic
  reasons.

We've already examined the second point and a reasonable amount of evidence for its validity has been presented. At this time, let's recall only one witness. Here's
Lynn Steen, a past president of The Mathematical Association of America[1]:

> Despite an objectivity that has no parallel in the world of art,
> the motivation and standards of creative mathematics are
> more like those of art than of science. Aesthetic judgements
> transcend both logic and applicability in the ranking of mathe-
> matical theorems: beauty and elegance have more to do with
> the value of a mathematical idea than does either strict truth or
> possible utility.

My first point receives support from Thomas Munro,
who opens his book *Toward Science in Aesthetics* with the
paragraph[2]:

> In spite of many attempts to turn it into a science, aesthetics is
> still a branch of speculative philosophy. Among all the
> branches of philosophy, it is probably the least influential and
> the least animated, although its subject matter—the arts and
> the types of experience related to them—is quite the opposite.

Arthur Berger, in the introduction of D. W. Prall's *Aesthetic
Analysis,* refers to J. A. Passmore's characterization of aesthetics as "an attempt to construct a subject where there
isn't one." Berger points out that Passmore's remark was

made in an essay with the provocative title "The Dreari-
ness of Aesthetics" and he suggests Passmore's attitude to
be "one which many of us secretly shared."[3]

Prall wrote in 1936 but the notion persists. The contemporary Yale philosopher Nicholas Wolterstorff has described aesthetics as being a "somewhat turgid backwater
alongside the main currents of philosophy."[4]

Many reasons exist for the second-class status of aesthetics among the areas of classical philosophy. They include the notion that art exists to be enjoyed rather than
analyzed and that, whatever the aesthetic theory which
emerges from the analysis, it will be so highly abstract and
conceptual as to have little to do with the art which it pur-
ports to explain. In addition, there exists the widespread
belief that aesthetics—whatever it may be—can be of little
importance. Questions regarding truth or justice or reality
may have value because these subjects have value. But
beauty and art, however carefully they are defined, are by
comparison frivolous concepts and unworthy of serious
consideration. Truth, justice, and reality, they say, are
proper philosophical objects but beauty and art are concepts for dilettantes. Edward Bullough, who made important contributions to aesthetics, spoke of the subject as his
"intellectual hobby." Art, you see, exists mainly to give
pleasure. And serious people—most of them—do not deal
professionally with matters of amusement and enjoyment.

But they ought. Because there is more at stake than
understanding what kind of art appeals to whom and why it
does. Much more needs to be explained than simple issues
of taste and distaste, of likes and dislikes. There is more to
be understood even than matters of art criticism, as difficult and as economically significant as these issues may be.
We need to understand why people are driven to do impor-

tant and difficult things for the sake of beauty and for its
sake alone. We need to understand what Mr. Keats had in
mind when he wrote[5]:

> "Beauty is truth, truth beauty,"—that is all
> Ye know on earth, and all ye need to know.

And we also need to know why a collection of strange
and peculiar people write 25,000 research papers each year
on an esoteric subject called mathematics.

THE DELICATE SIEVE

Deeply embedded in our culture lies the notion that
mathematics can be truly comprehended only by a gifted
minority. Because so few members of the otherwise educated public possess even the rudiments of mathematical
knowledge, mathematics has been assigned a special status.
Unique among the collection of disciplines—such as philosophy, history, and literature—which in times past
formed the basis of both the concepts of liberal education
and the core curriculum, mathematics may be *set aside*
under ordinary conditions without social or intellectual
consequence. People who would be loath to admit their
ignorance of classical music or of the contemporary novel
do not hesitate to announce their mathematical deficien-
cies. The phrase: "I never was any good at mathematics"
recurs like Malibu surf. And you hear it as often from concert goers and museum patrons as you do from gas station
attendants who take your money and try to count out your
change.

Behind all this stands the concept—widely held and
deeply believed—that there exists in a certain few
members of the human race a type of "mathematical

mind" which *allows* them to understand the logical complexities of mathematics. It is believed that just as there are only a few people capable of running 100 meters in less than ten seconds there are only an analogous few capable of *understanding* mathematics. And just as the inability to sprint at world-class level carries with it no social stigma, neither does the inability to understand mathematics.

Mathematical talent comes to you exactly as does sprinting talent: God either gives it to you or he does not. Or so it is believed.

Such beliefs provide comfort. Through them, members of the public can justify their often awesome mathematical ignorance. And because of them, mathematicians can rationalize their failure to teach—in a way such that the knowledge sticks—even the basics of the glorious discipline which occupies every moment of their conscious thought and almost every ounce of their energy. You cannot be expected to understand mathematics—so goes the myth—unless nature has provided to you the kind of mind necessary for the subject's comprehension. Nor can you be expected to teach mathematics to people who lack the basic mental equipment for it, just as Minnesota Fats lacks the physical equipment to run in the Olympic Trials.

Comforting they may be, but these beliefs have no more validity than astrology. They no more explain the dearth of common mathematical knowledge or the failure of mathematics teaching than the 1960's antisocial behavior has been explained away by labelling it "student unrest." Naming a thing does not necessarily help you understand it. Nothing is gained by learning that you suffer from paranoia if others are really out to get you. Attributing teaching and learning failure to something called "math anxiety" serves no purpose except to provide a built-in excuse for inadequate performance on both sides.

In the first place, what we expect our students to learn about mathematics is not much. In the university, we see the deterioration particularly with respect to calculus instruction. We've seen how this potentially lovely and fundamental course has been reduced to a grab bag of techniques and applications which—because they lack depth and content—serve no purpose other than to widen the gap between those who truly understand and appreciate mathematics and those who do not. Critic after critic has called attention to the meager intellectual content of calculus *as presently taught* by a generation of mathematicians who have essentially given up on the notion that the depth and subtleties of the subject can be communicated.

Moreover, we have concluded that mathematicians mainly pursue mathematics for the aesthetic experience it provides. And each of us who has ever sat in any undergraduate mathematics classroom knows from painful experience that the professor kept the aesthetic side of mathematics away from us just as Mrs. Robinson kept secrets away from the kids. Certainly, we shall never know whether or not mathematics can be made intelligible to the general public until some serious and systematic attempt has been made to present the subject at the elementary levels as the mathematicians see it at the highest levels: as the pinnacle of the arts, a thing as lovely and seductive as Cleopatra.

It may be that in the final analysis what distinguishes the mathematical mind has little to do with logic or precision or the ability to manipulate algebraic symbols or even the proficiency to deal with layers of abstraction. One person who saw this most clearly was the French mathematician and philosopher Jules-Henri Poincaré. Poincaré—whose gigantic output of writings was intended not only for mathematicians and philosophers but for educated people in all walks of life—believed that an "aesthetic sensibility" for mathematics defined the very soul of the mathematician. It acted as a "delicate sieve" without which no one in mathematics can become a "real creator."

A FURTHER STEP

I have no doubt that Poincaré was correct. In the "Mathematical Creation" chapter of *Science and Method,* Poincaré argued that facility with logic and the ability to manipulate mathematical symbols are insufficient to enable one to *create* mathematics. What is needed—and what all research mathematicians have deep inside them—is an "intuition" which allows both the conscious and the subliminal mind to sort out, from the myriad of mathematical possibilities before it, those which possess the "character of beauty and elegance and which are capable of developing in us a sort of aesthetic emotion." He wrote[6]:

> To create consists precisely in not making useless combinations and in making those which are useful and which are only a small minority. Invention is discernment, choice. . . . The useful combinations are precisely the most beautiful, I mean those best able to charm this special sensibility that all mathematicians know, but of which the profane are so ignorant as often to be tempted to smile at it.

[Incidentally, Poincaré's use of the word "intuition" agrees with its common meaning except for time scale. Ordinarily, we mean "intuition" to refer not only to an extralogical mental process but also to one which occurs quickly. We may think of an intuitive answer as an answer arriving from an almost instantaneous, nondeductive thought process. Mathematical intuition—as Poincaré intends it— lives outside ordinary reason in the usual sense but it may come about slowly. The subliminal mind may not work fast. Instant, intuitive guesses about mathematics, even by mathematicians—are often wrong.

At one time or another, my own quick intuition about each of the following questions was dead wrong. You might want to amuse yourself by trying your own hand. Just read each question, give it no more than a minute's thought, and mark it *true* (T) or *false* (F). Answers are provided at the end of the chapter.

1. _____ A small village has a single male barber. None of the male inhabitants of the village wear beards. The barber shaves each man who does not shave himself. It follows that the barber shaves himself.

2. _____ In No. 1 it follows that the barber does not shave himself.

3. _____ Fifty people are stopped at random and asked their birthdays. It is unlikely that any two of them were born on the same day of the year.

4. _____ A skier stands at a point on the mountain. When he skies from this point in any straight line he goes downhill. It follows that he is standing on a peak.

5. _____ A couple has two children. One of the children is a girl. The probability that both are girls is one-half.

6. _____ Adams, Brown, and Carter fight a triangular duel by shooting at one another in cyclic order until two are dead. At each turn the shooter may fire any way he pleases. Adams never misses, Brown is 80 percent accurate, and Carter is 50 percent accurate. Carter has the first shot. His best strategy is to fire at Adams.

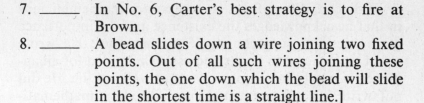

7. _____   In No. 6, Carter's best strategy is to fire at Brown.

8. _____   A bead slides down a wire joining two fixed points. Out of all such wires joining these points, the one down which the bead will slide in the shortest time is a straight line.]

Poincaré is right as rain about beauty and every mathematician knows he is right. You have inside you an aesthetic sensibility toward mathematics which acts on your intuitive mind as a delicate sieve sorting out the elegant and harmonious ideas from those which are merely useless combinations of other ideas. You have it, that is, or you are not a mathematician.

Unfortunately, however, Poincaré believes this sensibility to be *innate* and that only a relative few possess it. He said[7]:

> We know that this feeling, this intuition of mathematical order, that makes us divine hidden harmonies and relations, can not be possessed by every one. Some will not have either this delicate feeling so difficult to define, or a strength of memory and attention beyond the ordinary, and they will be absolutely incapable of understanding higher mathematics. Such are the majority.

Poincaré wrote these lines almost one century ago. In his world and with his view the profane formed the vast majority, and sensitivity to the aesthetics of mathematics was something granted only to the tiny subset of the population destined to be creative mathematicians. In holding this view, Poincaré falls right in line with the unstated but tacitly held position of contemporary mathematicians who also believe this aesthetic sensibility to be innate. If you have it, they believe, you will become one of them. If you do not, nothing can be done for you.

Poincaré differs from contemporary mathematicians in that he acknowledges the existence and the importance of the aesthetic component of mathematics. He wrote about the concept in books and essays intended for an audience of intelligent people in various walks of life. He did not write just for mathematicians. Modern mathematicians speak of this aspect of their subjects only in whispers, and only to one another. Poincaré tells the aesthetic story to all of us.

But perhaps Mr. Poincaré did not go far enough. He asserted (correctly) that the distinguishing feature of the mathematical mind is not logical but rather is aesthetical. But he is speaking of mathematics at the highest levels and when he referred to the "true aesthetic feeling that all real mathematicians know" he is talking about the *creation* of mathematics. And by "real mathematicians" he meant "research mathematicians." He does not claim that either the understanding or the teaching of elementary mathematics could be enhanced by developing in students an appreciation for the aesthetics of mathematics. When Poincaré writes of the significance of the "aesthetic sensibility" as related to mathematics, he didn't have teaching in mind.

But maybe he should have. Perhaps Poincaré was wrong. Maybe the notion extends. Perhaps you can bring students to mathematics early on by emphasizing its aesthetic value rather than its utility or its applicability. Maybe the poets can be led to learn mathematics if they are shown that there is poetry in the subject. Maybe the humanists will stay with mathematics a bit beyond the required minimum if they have developed within them what Poincaré called "our natural feeling for mathematical elegance."

I don't know. But one thing is certain. We will do no harm by trying. For what we do now has failed. Mathematics as understood by mathematicians remains unknown to everyone else. To the working scientist, it has the same status as the equipment bound in cold metal on his office shelves which he uses day to day in his work. To the humanist—and to everyone else in the other culture—the subject is something to be shunned.

I refuse to believe that this is the nature of things, that mathematics must remain forever beyond all but a tiny minority of our citizens. The notion that there exists a large subset of the populace who are capable of appreciating and understanding music, art, and literature but are somehow innate mathematical cripples seems to be simultaneously arrogant, apologetic, and just plain wrong. But, obviously, we have not reached these people with our present system of mathematics instruction which turns on the concept that mathematics is best presented through emphasis on its value as a scientific tool. We can do ourselves no harm by trying another approach, by presenting to our students early on those characteristics of mathematics which, in Poincaré's words, contain "this character of beauty and elegance, and which are capable of developing in us a sort of aesthetic emotion."[8]

AN EXAMPLE

In 1978, Seymore A. Papert of MIT suggested that Poincaré's notion might be taken a further step into the teaching of mathematics. In an essay called "The Mathematical Unconscious," Papert considers the possibility of extending Poincaré's ideas—"bringing the theory down to earth"—in such a way as to make it applicable to the teaching and the learning of mathematics. Papert's idea consists mainly of the observation that Poincaré could be right about the significant connection between mathematics and aesthetics but wrong about the recognition of this connection—Poincaré's "aesthetic sensibility"—being innate and incapable of being learned. Papert writes[9]:

> The destructive consequences of contemporary mathematics teaching can also be seen as a minor paradox for Poincaré. The fact that our schools, and our culture generally, are so far from being nurturant of nascent mathematical aesthetic sense in children causes Poincaré's major thesis about the importance of aesthetics to undermine his grounds for believing in his minor thesis which asserts the innateness of such sensibilities. If Poincaré is right about aesthetics, it becomes only too easy to see how the apparent rareness of mathematical talent could be explained without appeal to innateness.

Sure. The mathematical elite share with Poincaré the sensitivity to the aesthetics of mathematics. Of this, there can be no doubt. But maybe the rest of us can share it too *provided the methods of presenting and writing mathematics are changed so as to bring the aesthetic component out into the open* and to develop a sensitivity for it in students as we now attempt to develop in them a sensitivity for the poetry of Shakespeare, the music of Brahms, and the paintings of Cézanne. We cannot *assume,* as Papert points out, that Poincaré's aesthetic notions (which he associated only with the highest levels of mathematics) apply to more elementary mathematical work. But, until we have some data and some experience with alternate approaches to mathematics, *we cannot assume the contrary either.* There may exist in the learning and doing of even elementary mathematics a previously unexplored concept of *aesthetic guidance.*

Papert suggests that the concept does exist. He describes a psychological/mathematical "episode" which partially supports his thesis. In the episode (an actual psychological experiment), a number of human subjects were asked to construct a proof of the famous and ancient result which caused the Pythagoreans so much dismay. The subjects were asked to prove that the square root of two is irrational.

Recall that a real number is called rational if it can be represented as the quotient of two integers. For example, 1/2, 2/3, and 76/89 are all rational numbers. An integer, 6 for example, is rational since it can be represented as itself divided by 1: $6 = 6/1$. The Pythagoreans, you will remember, believed that nature itself could be reduced to integers and the quotients of integers. They were much shaken by their own discovery that the number x which satisfies the equation $x^2 = 2$ is not rational.

Numbers which are not rational, i.e., which cannot be represented as the quotient of two integers, are called irrational. The number x whose square is 2 is called the "square root of 2" and is denoted by the symbol $\sqrt{2}$. In the episode described by Seymore Papert, subjects of varying levels of mathematical sophistication were asked to produce a proof of the following:

Theorem: $\sqrt{2}$ is irrational.

This theorem and its standard proof are both justly famous. In *A Mathematician's Apology,* G. H. Hardy presents this result as one of two "real mathematical theorems, theorems which every mathematician will admit to being first rate" (the other theorem is Euclid's 300 B.C. proof that there exist infinitely many primes). And Hardy says of the two famous theorems of Greek mathematics[10]:

> They are "simple" theorems, simple both in idea and in execution, but there is no doubt at all about their being theorems of the highest class. Each is as fresh and significant as when it was discovered—two thousand years have not written a wrinkle on either of them.

Highest-class indeed—and Hardy presents them as examples of "serious mathematics" because, as he writes,

> The beauty of a mathematical theorem depends a great deal on its seriousness, as even in poetry the beauty of a line may depend to some extent on the significance of the ideas which it contains. . . . Beauty is the first test: there is no place in the world for ugly mathematics.

Let's return to our theorem and its proof.

Theorem: $\sqrt{2}$ is irrational.

Proof: Suppose the theorem is false; i.e., suppose $\sqrt{2}$ is rational. Then we may write

$$\sqrt{2} = \frac{p}{q}$$

where p and q have no common factors. (Any original common factors may be canceled leaving numerator and denominator free of them.) Hence,

$$\sqrt{2}\, q = p$$

or

$$2q^2 = p^2.$$

Thus, p^2 is an even number. It then follows that p is an even number. (It is easy to see that the square of any odd number is odd.) Thus, $p = 2c$ for some integer c. Therefore,

$$2q^2 = (2c)^2$$

or

$$q^2 = 2c^2.$$

Thus, q^2 is even and, just as before, it follows that q is even. Consequently, both p and q are even numbers and so are both divisible by 2. This contradicts our assumption that p and q have no common factors. So our hypothesis that $\sqrt{2}$ is rational is false. So, $\sqrt{2}$ is irrational and the theorem is proved.

Let us consider this theorem and its proof from the standpoint of beauty and elegance. In his *Apology*, Hardy tries to come to terms with some of the characteristics which give the result its elegance. Among the items which Hardy mentions are *seriousness, depth, generality, unexpectedness, inevitability,* and *economy.* Viewed appropriately, there can be no doubt that the theorem and its proof fulfill each of the conditions set by Mr. Hardy. However, it is likely that the ordinary meanings of Hardy's terms do not immediately carry over to mathematics and you may not see instantaneously and independently that the theorem and the proof are aesthetically pleasing. One of my quarrels with the present system of teaching mathematics in the United States results precisely from the fact that—outside the aristocracy of mathematicians—no one has been trained to look at mathematics as art. Consequently, whenever you offer up for the first time a mathematical result and point to it as an example of something of aesthetic value, people outside mathematics—even those with considerable knowledge and experience of the "ordinary" arts —may fail to see what the fuss is all about. In this case, otherwise arts-knowledgeable people become analogous to children raised on rock music who, for the first time, are brought inside a concert hall and exposed to the Beethoven Violin Concerto. They cannot be expected to recognize immediately the concerto as great art. Nevertheless, if they are educated and intelligent they can be expected to believe in the aesthetic value of the piece because they know of the existence of 150 years of performance and criticism which have minutely examined the concerto and confirmed its greatness. *Belief* in the concerto's value comes first to the student and brings him inside the concert hall. Subsequent *experience* and *knowledge* allow the student ultimately to evaluate the concerto for himself. When he does, we will all be surprised if he finds it trivial, shallow, specialized, expected, avoidable, or slack—or describable by any other collection of adjectives which are the opposites of Hardy's criteria for mathematical elegance.

Similarly, you should be willing to accept for the moment the aesthetic value of the proof of the irrationality of $\sqrt{2}$, not because I say so but, rather, because a great mathematician named G. H. Hardy does so and because a long line of mathematicians from the time of Pythagoras to the present have concurred. (What is missing here, of course, is the existence of an established culture and criticism analogous to that of music which will enable you ultimately to decide the mathematics aesthetic matter for yourself. There exists an accessible literature which you can consult in order to determine the relative merits of the music of, say, Tchaikovsky and Stravinsky. No such critical literature exists for mathematics. But that is what this book is about and it is the main point of my call for reform in the teaching and learning of mathematics. Intelligent people need to be able to see mathematics as the mathematicians see it. They need not learn to do mathematics as the mathematicians do it. But, in order to see, they need experience and training in how to look at mathematics just as they need experience and training in how to listen to Beethoven. The natural place for this instruction is in the mathematics classroom.)

Expressed very briefly, Poincaré's idea consists of his determination that mathematical creativity involves both the conscious and the unconscious mind and that the creative process falls into three distinct pieces; a first stage involving conscious analysis, a second stage of unconscious work, and a final stage in which the product of the unconscious passes back to the conscious mind. If the problem is of sufficient difficulty, Poincaré believes it could not be solved in stage one. Stage two, he says, has to occur even if it takes place after the mathematician believes the problem has been abandoned. Finally, the result of the unconscious pops back into the conscious mind at seemingly random times which are usually unrelated to what the mathematician may be doing or thinking at the moment of revelation.

The unconscious mind, Poincaré asserts, cannot rigorously determine mathematical solutions or necessarily bring forth only correct results. However, the ideas brought forth from the unconscious and handed over to the conscious invariably possess the stamp of mathematical beauty. Furthermore, there seems to be at work a high and mystical aesthetic principle which produces a positive correlation between the elegance of the mathematical idea and its correctness and importance—or, as Hardy put it, a positive correlation between the beauty of the idea and its *seriousness.*

In "The Mathematical Unconscious," Seymore Papert speculates on whether or not Poincaré's aesthetic monitor might also work in the conscious mind and whether it might also operate at less lofty mathematical levels. Does there exist, Papert wonders, a principle of aesthetic guidance which will help nonmathematicians to learn or to re-create mathematics while they are actively engaged in the process of learning or doing. If so, then the potential effects of replacing even part of the present failed system of mathematics teaching with a new approach involving the utilization of the principle of aesthetic guidance are obvious and far-reaching. Papert puts it this way[12]:

> Bringing his theory down to earth in this way possibly runs the risk of abandoning what Poincaré himself considered to be most important. But it makes the theory more immediately relevant, perhaps even quite urgent, for psychologists, educators, and others. For example, if Poincaré's model turned out to contain elements of a true account of ordinary mathematical thinking, it could follow that mathematical education as practiced today is totally misguided and even self-defeating. If mathematical aesthetics gets any attention in the schools, it is as an epiphenomenon, an icing on the mathematical cake, rather than as a driving force which makes mathematical thinking function. Certainly, the widely practiced theories of the psychology of mathematical development (such as Piaget's) totally ignore the aesthetic, or even the intuitive, and concentrate on structural analysis of the logical facet of mathematical thought.

Yes. Mathematical aesthetics of the schools as an *epiphenomenon.* At most it is this. If, indeed it is anything at all. Instead of aesthetics, the schools give us drill and tedium—and immediately forgettable techniques aimed at unwanted and unwelcome applications.

In the "episode" described by Papert, a number of subjects were given the equation

$$\sqrt{2} = \frac{p}{q} \text{ , where } p \text{ and } q \text{ are integers,}$$

and asked to show that it could not hold by demonstrating that it implied something bizarre or contradictory. In other words, they were asked to construct a proof of the theorem asserting the irrationality of $\sqrt{2}$. Papert writes[13]:

> Almost as if they had read Freud, many subjects engage in a process of mathematical "free association" trying in turn various transformations associated with equations of this sort. Those who are more sophisticated mathematically need a smaller number of tries, but none of the subjects seemed to be guided by a provision of where the work will go. Here are some examples of transformations in the order they were produced by one subject:
>
> $$\sqrt{2} = p/q$$
>
> $$\sqrt{2}q = p$$
>
> $$p = \sqrt{2}q$$
>
> $$(\sqrt{2})^2 = (p/q)^2$$
>
> $$2 = p^2/q^2$$
>
> $$p^2 = 2q^2.$$

As we have seen in our proof of the theorem, the key equation in the subject's list is the last one. From it one concludes that p^2, and hence p, is an even number. It is then a natural step to deduce that q also must be even, so that the original equation cannot hold for whole numbers without common factors. But, if it holds at all, then common factors could be canceled. Consequently, the original equation cannot be valid.

The equations in the above list are all mathematically equivalent in that they contain the same information and any one of them can be easily obtained from the other. The last equation, however, is the one you want if you are to prove the theorem. Does anything set it apart from the others? How do you recognize the significance of this equation when it appears embedded in a string of relations all yielding the same mathematical information. Here is Papert again[14]:

> All subjects who have become more than superficially involved in the problem show unmistakable signs of excitement and pleasure when they hit on the last equation. The pleasure is not dependent on knowing (at least consciously) where the process is leading. It happens before the subjects are able to say what they will do next, and in fact, it happens even in cases where no further progress is made at all. And the reaction to $p^2 = 2q^2$ is not merely affective: once this has been seen, the subjects scarcely ever look back at any of the earlier transforms or even at the original equation. There is something very special about $p^2 = 2q^2$. What is it?

Mr. Papert is too wise and too experienced to provide an unequivocal answer to his question. But he thinks he knows. And so do I. The equation brings pleasure and excitement because it has an aesthetic value that the others do not. Papert attempts to analyze the aesthetics of the situation by arguing that $p^2 = 2q^2$ "resonates" with other items or processes which are accessible to the conscious mind. He suggests that, while the original equation setting up the problem focuses on 2 and relegates p and q to "dummy" roles, the final equation turns p into a "subject" and causes the original subject, $\sqrt{2}$, to vanish completely. This, he says, provides a situation "as sharply different as in the figure/ground reversal or the replacement of a screen by a face in an infant's perception of peek-a-boo."[15]

Maybe. But peek-a-boo or not, it is clear that Papert is on to something. For, as he says, "the mathematics of a mathematician is profoundly personal" and yet, when mathematics is taught in the schools, students are asked to "forget the natural experience of mathematics in order to learn a new set of rules."[16] Moreover, as we all know, the existing "rule learning process" does not work, has not worked, and—in my view—cannot work.

What is needed is a real understanding of the mathematician's "personal experience" with his subject. At the highest levels, there can be no doubt that this experience is largely aesthetic. What we must learn first are the characteristics of mathematical aesthetics so that we can talk about mathematical elegance in more than a descriptive manner. Then we must determine whether or not Poincaré's principle of aesthetic guidance can be brought down to earth to aid in the process of the teaching and learning of elementary mathematics.

The first task will require a rethinking of classical aesthetics with the aim of developing theories which will apply to mathematics as they now do to the traditional arts such as music, literature, sculpture, and painting. When the professor asks the students in Philosophy 101, "What is beauty?" it must be clear that he includes mathematics as one of the question's objects. The second chore will necessitate the revamping of the present system of mathematics instruction. In the schools, it will require a new generation of mathematics teachers: people who themselves have been personally touched by mathematics deeply enough to have some chance at communicating to their students a semblance of the excitement of the subject. In the colleges, it will mean that the research mathematicians will have to return to teaching in the sense that this activity becomes once more a priority and in such a manner that the fragmented, hit-or-miss, grab bag of techniques which make up the basic curriculum is abandoned in favor of courses filled with concepts and ideas. The concepts of mathematics must be motivated for the students as they are motivated for the mathematician—by the beauty and elegance of the subject, the loveliest subject on the face of the earth.

THEORIES

Of the two tasks just mentioned, the easiest—from a theoretical point of view—is the second. At the college level, in fact, there are no theoretical problems connected with the return of mathematicians to teaching or to the rewriting of the curriculum along the lines of mathematical concepts motivated by aesthetics. The mathematicians themselves deal with ideas and they themselves are motivated by aesthetics. They can—if they wish—bring these concepts and this motivation to the calculus classroom tomorrow.

The trouble, of course, is they do *not* wish. Life at the top of the mathematical chain is now—and has been for twenty-five years—relatively pleasant. Mathematicians in the major universities *do* mathematics without the bother of having to become deeply involved in matters associated with teaching or curriculum. Although you might easily convince them one-on-one that concepts need to be refitted to the basic mathematics curriculum replacing the hodge-podge syllabi of the past two decades, you will be hard pressed to bring forward any *organized* movement of mathematicians advocating such reform. For an organized effort would take the mathematicians' time and energy away from the creation of mathematics which they do well and require their devotion to the practice of real-world academic politics and negotiation which they do not at all.

Nevertheless, reform is possible in the sense that there exist no *theoretical* obstacles to bringing it about.

The aesthetic theory question, however, is another matter entirely. Not that there is a paucity of aesthetic theories—philosophical libraries teem with them. Plato, Aristotle, Kant, Croce, Collingwood, Langer, Whitehead, Goodman, Danto, Dickie, and many other philosophers both classical and modern have dealt with the question of the definition of beauty and the quest for the characteristics of art. But almost nowhere in their confusing array of theories and countertheories can one easily find a body of organized knowledge which helps with the determination of exactly what the mathematicians have in mind, consciously or unconsciously, when they describe—with unanimous and easy agreement—a mathematical result as "elegant."

You can find a number of works of varying degrees of seriousness and intelligibility on the general subject of mathematics and art. These are books and articles which examine, in one way or another, the role of mathematics as *applied* to the arts as the arts are ordinarily defined. In the past, for example, these studies have included endless speculations on such topics as the golden rectangle (a rectangle in which the ratio of lengths of sides approximately equals 1/1.618), its appearance in the classical paintings of the great masters and its occurrence in ancient architecture like that of the pyramids or the buildings and bridges of Greece and Rome. More recently these studies have included work on symmetry in art by Hermann Weyl,[17] and mathematics and music by Sir James Jeans.[18] We will have occasion to mention Gustav Fechner,[19] who attempted in 1876 to produce "scientific experiments" in the field of psychology which included, in particular, a paper on the golden rectangle. Fechner's influence continues up to the present day and we see it reflected in contemporary psychological research like that of McManus, Edmondson, and Rodger,[20] who reported on the phenomenon of "balance in pictures"—the viewer's subjective feeling associated with a notion which ought somehow to be described in mathematical terms of symmetry or center of gravity.

There exists much, much more work along these general lines which, essentially, constitutes a kind of nonrigorous application of mathematics to the arts. This research may, in fact, be important. The golden rectangle is a legitimate mathematical object and its interesting properties appear in unexpected places such as quotients of terms of the Fibonacci sequence and in the radius of convergence of certain power series. The fact that the front of the Parthenon at Athens (with its ruined pediment added on) fits almost exactly into a golden rectangle may tell us something meaningful about the mind-set of the Greek architectural masters. And the fact that the golden rectangle appears regularly in the paintings of Da Vinci and Seurat no doubt is also significant.

Moreover, there exists another, albeit thinner, body of work connecting mathematics and aesthetics. In this direction, researchers have tried to apply mathematics to the concept of aesthetics itself, thus bringing mathematical concepts and methods to the classical philosophical question of the recognition of beauty. This research includes "Mathematics of Aesthetics" by the mathematician George David Birkhoff,[21] and *Algorithmic Aesthetics* by George Stiny and James Gips.[22] Birkhoff attempts to describe a significant portion of aesthetic theory with mathematical formulas. He identifies three typical "aesthetic experience" variables: the *complexity* (C) of the art object, the harmony or *order* (O) of the object, and the *aesthetic measure* (M) of the object. Birkhoff asserts these variables are related by the basic formula

$$M = \frac{O}{C}$$

and he produces a number of examples to support his claim. Stiny and Gips, on the other hand, are concerned with developing and examining explicit, step-by-step procedures—called algorithms—which can be described in mathematical terms and which, when properly posed, will produce methods of aesthetic criticism straightforward enough to be handled by a computer.

Each of these areas of research involving mathematics and aesthetics—the application of mathematics to art and the use of mathematics to produce aesthetic theories—has its own charm and, no doubt, its own real value. But neither is helpful to us here. Our problem is *to examine mathematics itself as art.* What we need are aesthetic theories which will address the specific question: What is *mathematical* beauty?

Alas, specific theories do not exist. Or, if they do, they are obscure enough so as not to exist for all practical purposes. As far as I can see, the philosophers who have dealt seriously with aesthetics are either uncomfortable with mathematics or else they have not considered the question of mathematical beauty to be significant. Or—what is more likely—many aestheticians, along with the other educated people who live outside the mathematical aristocracy, have no notion that the concept of mathematical elegance or mathematical beauty even exists. For they will have had the same mathematical education as have the rest of us. Consequently, they will no more believe mathematics can be lovely than they will believe the sea is boiling hot, or that pigs have wings.

And the mathematicians, who instinctively—and according to Poincaré, innately—*understand* the mathematical aesthetic experience, have no apparent interest in *analyzing* it. Like the farmer who knows nothing about art but knows what he likes, the mathematicians know elegance when they see it. They see no need to explain it.

ANALYTIC PHILOSOPHY

We must face the possibility that the absence of a precise aesthetic theory may simply be in the nature of things. More precisely, the following may hold:

(*) A true aesthetic theory which includes mathematics cannot exist.

We want to look at statement (*) closely but first let's note that there are contemporary philosophers who doubt that any grand aesthetic theories are possible. Two who have expressed recent doubt are Mortimer Adler and George Dickie.

In *Six Great Ideas,* Adler[23] tells us that while the philosopher can be "intellectually responsible" in the areas of truth and goodness, he is unable—presumably because of the nature of things—to discharge his responsibility in matters regarding the "idea of beauty." A reason for this, Adler says, is that the objectivity associated with truth and the *near* objectivity of goodness gives these concepts "values" which do not translate to the notion of beauty which necessarily depends in some way on the "taste" of the observer.

In his 1984 book, *The Art Circle,* George Dickie[24] asserts his own doubt that any theory can be "worked out" to account for (what he calls) aesthetically relevant kinds of properties. The best that can be done, Dickie thinks, is to deal piecemeal with particular works of art one-by-one.

Adler and Dickie are expressing informed doubt about the existence of certain types of aesthetic theories and it is well that we take them seriously. However, we may be able to say more about aesthetic theories which include mathematics. There may be a theoretical reason for the validity of statement (*) or, at least, for a weaker version which says that any existing theory cannot in the long run be helpful.

I must digress to make this clear. If the work "theory" in statement (*) is interpreted strictly, then by it we mean a framework or set of rules which prescribes the collection of *allowable* sentences one can make about whatever objects are under consideration. For example, the theory of quantum mechanics in physics leads to a result called the Heisenberg Uncertainty Principle. This principle, a *consequence* of the theory, says that one cannot know *both* what a fundamental particle is doing and where it is located. That is, an electron (for example) does not possess simultaneous momentum and position. Under the assumptions of quantum mechanics, you simply cannot talk about this kind of simultaneity. Any sentence which does is not allowable.

A theory, therefore, gives you a framework for talking about the objects of the theory. More precisely, it gives you rules of *inference* from which you can prove theorems about the objects. *Truth* about the objects consists of sentences about them which are either assumed in the theory or else are provable within the framework of the theory.

For example, a truth about a moving particle in the theory of Newtonian mechanics is that it will continue to move along a straight line unless it is acted on by some outside force. This "truth" is roughly Newton's First Law of Motion and is a basic assumption of his theory. Aristotle, on the other hand, said something entirely different, namely, that a moving object would continue to move only so long as a force is applied to keep it moving.

Which statement is true? Newton or Aristotle? The answer depends on which theory you are using. (Most college physics students would side with Newton since his first law forms a fundamental part of what they study. On the other hand, Aristotle's idea agrees more closely with our everyday experience. Anyone who has tried to push a stalled car along a level road knows what happens when he stops shoving.)

In a certain real sense, all true theories are *mathematical theories*. (By a true theory I mean a rigorous theory containing axioms, definitions, and rules of inference.) What you do is replace the real objects by mathematical counterparts and the assumptions of the theory by mathematical statements (usually sets of equations). The rules of inference which are allowable then become the rules and logic of mathematics.

Obviously, this concept of "mathematical theory" fits neatly with our earlier discussion of the applied mathematics process. In fact, the two concepts essentially coincide. All you need do is look back at Figure 7 and replace the real-world rectangle by another which now represents those objects you want to examine. The rest of Figure 7 remains unchanged except that you now think of the mathematical model of the objects as being the "theory" of the objects. You discover "truth" in the mathematical world by the analysis process and you "apply" this truth back to the object world by the application process of the diagram (see Figure 10).

Ordinarily, you want a theory about real-world objects so that the object world in Figure 10 remains the real world, as Figure 7 shows, or a proper subset of the real world. But

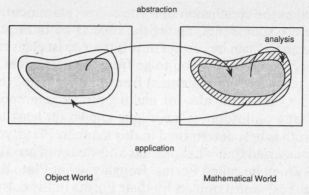

abstraction

analysis

application

Object World   Mathematical World

FIGURE 10. Mathematical theory.

if you wanted, for example, a mathematical theory of acting, then the object world might well include all past performances of Laurence Olivier at the Old Vic in London. (Whether such things do or do not live in the real world is a matter we, fortunately, need not consider at this time.)

So, in the case of a mathematical theory "truth" about objects means mathematical statements that can be deduced about them in the analysis stage of Figure 10. In the case of *any* theory—mathematical or not—truth is a property of what the theory allows you to say about the objects rather than a property of the objects themselves. (For example, in Newton's theory one can speak of simultaneous events, say two lightning flashes. In the Einstein theory, simultancity depends on the observer, not on the events. So, whether or not the flashes occur at the same time is a property of the theory, not of the flashes.)

At first glance, this notion of truth seems technical and sterile. However, if you pursue it far enough—as mathematicians (in particular) do each day—you will find it neither of these. Obviously, truth—whatever it may mean—will be of no value unless it can somehow be expressed as the conclusion of some correct statement, or set of correct statements, about the objects of interest. And "correctness" can be certain only if there exist clear rules of inference which allow you to go from one statement to another, to *deduce* one statement from another. The "theory" provides the set of rules for making proper inferences.

Many philosophers have long known (at least tacitly) that truth is best determined in this manner. Plato, you will recall, asserted that what one has knowledge of are abstractions, which he called Forms. Roughly, what Plato did was replace real-world objects by their Forms in his epistemology. Each real-world object—say an apple—is a temporal thing which undergoes constant change. The *Form* of the apple, however, lives in an ideal realm and is eternal and unchanging. It is the Form of the apple and not the apple, said Plato, about which we can have knowledge. Moreover, because Plato's truth could be known only about Forms and not about real-world objects the Forms were to him *more real* than the actual objects.

Thus, Plato's epistemology represents a first step toward the establishment of a "theory" of the type we are discussing. Had he embedded his forms in the mathematical world (and Plato knew much mathematics) his theory of knowledge would fit exactly into our diagram in Figure 10. He did not develop a mathematical epistemology, however, nor did he list a set of inference rules for dealing with statements about Forms. The resulting philosophy is therefore less exact but undoubtedly less restrictive and much more encompassing. Plato and the other classical philosophers proceeded—as Will Durant put it—"synthetically" rather than "analytically." Analysis belongs to science. "Science gives us knowledge, but only philosophy can give us wisdom."[25]

Philosophy has now become deeply analytical and

many of its critics, like Durant decades ago and Allan Bloom today, criticize it for exactly this reason. Durant wanted wisdom from philosophy and he wanted it "clothed in the living form of genius." He wanted wisdom from a subject which included the study not only of philosophy itself but of those philosophers who produced it. Durant wanted philosophy as truly a part of the humanities. Analysis, Durant claimed, had "dismembered" the subject and left it in "shrivelled abstractness and formality."

Allan Bloom's criticism is much the same. He longs for philosophy "not as a doctrine but a way of life" and asserts that the academy, in particular, depends essentially on this lifestyle. "The tiny band of men," Bloom wrote, "who participate in this way of life are the soul of the university."[26]

Maybe. Certainly to do analysis is to restrict yourself because the objects of inquiry are bound by the theory and the statements you can make about them are constrained by the axioms, the definitions, and the rules of inference. On the other hand, if you are looking for truth this is how you will find it.

Of course, the truth you find is truth about the objects of the theory. In the case of a mathematical theory these objects are mathematical objects and do not live in the world. In Plato's philosophy, the objects are Forms and are not things in the sense that apples or chairs are things. On the other hand, you should not leap to the conclusion that the objects of the theory are not *real* in the meaningful sense of the word. Plato argued, in fact, that Forms represented the only true reality since it was only Forms about which one could have knowledge. To a mathematician, the number 6 is a real thing even though it does not exist anywhere in nature. "Six" is a precisely defined mathematical object and about it certain truths are known. (For example, $6 = 1 + 2 + 3$; that is, "The integer 6 is the sum of its proper divisors." Such integers are called *perfect*. Six is the smallest perfect number. The next perfect number is 28. I do not know the third one. Nobody knows how many there are.)

It was precisely Plato's belief in the "reality" of his Forms which led him to his low opinion of art. To Plato, art meant imitation and, moreover, imitation of real-world objects. Since Plato regarded Forms as worthier than real-world things (because they were eternal, unchanging, and the objects about which truth could be known), he considered imitation a defect. For a picture of an apple is less real than the apple itself which in turn is less real than the Form of the apple. Consequently, art as imitation lived two levels removed from Platonic reality. (One wonders what Plato would have made of Impressionism and Postimpressionism. Perhaps one of Monet's bridges could be interpreted not as an imitation of a bridge, but rather as a representation of the Form of bridge, just as a mathematical model of a falling stone is a representation of the form of the stone provided by the theory of Newtonian mechanics.)

Although Plato viewed imitation as a defect, his authority set the Imitation Theory of Art in place for 2000 years. For all of this time, most art was constructed and evaluated in terms of the imitation theory—how well it represented the original object. (The theory remains forceful enough to bring discomfort to some aestheticians today. Norman Bryson[27] of King's College, Cambridge, argues in his 1983 book, *Vision and Painting,* that the pervasiveness of the "realist" view prevents meaningful change within the history of art and, moreover, that the notion that Western painting is an art of copying visual experience is itself a "radical misconception.")

Certainly, philosophy has in the past century turned more and more toward "analytic philosophy." By this I mean a movement toward what we have described as "theory," as the subject's practitioners wrestled ever more deeply with the meaning of the words they use and the way they fit them into sentences. Following Plato, Western philosophers devoted more than two millenia to a futile search for "absolute" truth—the futility being inevitable because of the heuristic and descriptive nature of their methods. In the twentieth century the analytic movement led by Bertrand Russell and Rudolf Carnap brought hard logic to philosophy. As we have seen, this process necessarily reduces the collection of sentences that can be used and the kind of inferences that can be made. As you move more closely toward analysis you tend to replace sweeping generalizations about the nature of truth or beauty by more technical declarations about the kind of questions that can even be addressed. Before you tackle "What is truth?" you must first make clear what the question means in light of your analytical framework. For example, the truth about simultaneous lightning bolts means one thing in Newtonian mechanics, but quite another in Einstein's physics.

This narrowing of the scope of philosophy has led to the kind of criticism we have observed in Durant and Bloom. Roger Scruton, Professor of Philosophy at the University of London, went so far with this criticism in 1983 as to assert: "It has been said, and with considerable truth, that contemporary analytical philosophy has deprived philosophy of its status as a humanity."[28] Let's hope Scruton is wrong. Philosophy outside humanities would be like mathematics apart from art.

What the analytical philosophers do is a kind of linguistic analysis. They attempt, as best they can, to make the language of their theory as precise as possible. But the language they want to use is natural language, not mathematics. They want to make their assertions in, let's say, plain English. But English has very complicated syntax, too complicated to allow complete mathematical modeling even if that were what the philosophers desired. Could they make such a model, their analysis would become part of symbolic logic which itself is part of mathematics. Particular analytic philosophical theory would then become a mathematical theory and the rules of inference would be the rules of mathematics. Then, and only then, would the theory—limited or not—obtain complete unambiguous precision.

At the moment, most of analytic philosophy falls way short of symbolic logic. Consequently, the theories fall short of mathematics and often seem—especially to a mathematician—as an odd blend of exposition, description, with a touch here and there of precision. But it is toward mathematical precision that the subject moves and its "theorems" approximate truth exactly as the theory approximates a mathematical theory. (As an indication of the direction in which analytic philosophy is moving I offer a conference announcement which crossed my desk yesterday. The conference is to be held at the University of Victoria in British Columbia by The Society for Exact Philosophy. The conference is called: "Automated Theorem Proving for Non-Classical Logics.) And if you are after "truth" about aesthetics you will find it in a mathematical theory of aesthetics, if such a theory can exist. (My own notion is that such a theory would be so limited as to be of little value. An interesting step toward such a theory can be found in the book *Algorithmic Aesthetics* by George Stiny and James Gips.)

Now let's return to statement (*) which claimed that a true aesthetic theory which includes mathematics cannot exist.

By the phrase "true aesthetic theory" I mean here an aesthetic theory through which we can make true assertions. That is, I mean the kind of theory toward which analytic philosophy tends and which it will reach once the appropriate modeling of language has been achieved. Thus, I mean a *mathematical theory*.

So, statement (*) now concerns itself with whether or not a *mathematical aesthetic theory* can exist which includes *mathematics*. The answer seems clearly negative. At least it is clear that no completely useful theory can exist. For consider the following. Suppose such a mathematical theory—call it T—exists. Consider a particular piece of mathematics, say Euclid's proof of the infinity of the collection of primes—denoted by p. A mathematician tells us that p is "beautiful." We want to check this by means of our aesthetic theory T. But the check itself, call it c_1, is also a piece of mathematics because T is a mathematical theory. Another mathematician, watching us bring c_1 to bear on p says: "c_1 is beautiful." We now use our theory T to check this claim and it does so by producing another piece of mathematics, c_2. The process now repeats itself indefinitely. "Is c_2 beautiful?" can only be answered by producing another check, c_3. Continuing, we obtain an arbitrary long sequence of pieces of mathematics, $c_1, c_2, \ldots, c_n$ with the property that our theory will check the beauty of the last one only by producing another. (Certain subtleties, like the possibility that all the c_i's might be identical, have been ignored in this heuristic analysis but the main point regarding the difficulties associated with such a theory has been illustrated.)

O.K., this tells us only that we must not look to mathematics for an aesthetic theory which *includes* mathematics. It does not mean that we should not look, but that we should look somewhere else. Look we must because before us remains the inescapable fact that mathematicians do mathematics for aesthetic reasons.

THE ARTWORLD

Arthur C. Danto is Johnsonian Professor of Philosophy at Columbia University and art critic of *The Nation.* Widely published, Danto has written more than ten books on such topics as philosophy, analytical philosophy, and art. On the paper cover of Danto's 1990 book, *Connections to the World,* Richard Rorty says: "Danto is as skillful and subtle as philosophers get."[29] High praise even if you restrict the comment only to Western philosophers and then only to those skilled as Plato, Aristotle, Spinoza, Kant, or Russell. And when Virginia's Richard Rorty speaks, people listen. Harold Bloom of Yale describes Rorty as "the most interesting philosopher in the world today."[30] Professor Danto, therefore, is someone you take seriously when you talk philosophy—maybe even someone to step aside from.

In 1964, Arthur Danto wrote a paper called "Artworld," which appeared in *The Journal of Philosophy.* Here, Danto impressively argues that works of art can exist only in the presence of some sort of artistic theory. He writes[31]:

> . . . one might not be aware he was on artistic terrain without an artistic theory to tell him so. And part of the reason for this lies in the fact that terrain is constituted artistic in view of artistic theories, so that one use of theories, in addition to helping us discriminate art from the rest, consists in making art possible.

This is not a commonplace observation and from it flows a pragmatic, if not completely logical, framework from which one can attempt to answer the ancient question: What constitutes a work of art?

As an example, Danto offers the advent of Postimpressionist paintings and argues that in the presence of only the Imitation Theory of Art, which had prevailed since Plato, they could not be considered art. He writes: "It was impossible to accept these as art unless inept art: otherwise they could be discounted as hoaxes, self-advertisements, or the visual counterparts of a madman's ravings." Only when a new artistic theory gained relevance, Danto claims, could these pictures truly become art. What was needed, and found, was an artistic theory which provided "a freshly opened area between real objects and real facsimiles of real objects." The new pictures then became "non-facsimiles" and "a new contribution to the world."[32]

A second example consists of a museum exhibit of Andy Warhol facsimiles of Brillo cartons, piled high, in neat stacks, as in the stockroom of the supermarket. What makes the Warhol exhibit art while the actual cartons in the supermarket down the street are not? Professor Danto answers[33]:

> What in the end makes the difference between a Brillo box and a work of art consisting of a Brillo box is a certain theory of art. It is the theory that takes it up into the world of art, and keeps it from collapsing into the real object which it is. . . . Of course, without the theory, one is unlikely to see it as art, and in order to see it as part of the artworld, one must have mastered a good deal of artistic theory as well as a considerable amount of the history of recent New York painting. It could not have been art fifty years ago. . . . The world has to be ready for certain things, the artworld no less than the real one. It is the role of artistic theories, these days as always, to make the artworld, and art, possible.

The artworld then represents some kind of institutional complex consisting of artistic theories and persons who have mastered the theories as well as a certain amount of art history. It is entrance to this artworld that gives an object the status of art. (Eugene F. Kaelin, Professor of Philosophy at Florida State University, compares this artworld entrance with that of "state of grace." He writes in the book *Aesthetics and Art Education:* "What makes art of an artifact is its entrance into the 'artworld,' the institutional complex within which bona-fide works of art receive the ascription of art much in the same way as a child becomes a 'Christian' at baptism."[34])

Let me point out in passing that Danto's "Artworld" paper is not easy reading. He devotes the final section to an analytical examination of the notion of "artistically relative predicates" which essentially are statements that, when they apply to an object, allow its entrance into the artworld. He develops a formal symbolic scheme for describing available artistic styles, given the active critical vocabulary: representational expressionistic, representational nonexpressionistic, etc. In this analysis Danto must, as he put it, "beg some of the hardest philosophical questions I know."

The paper also provides an example of the analytic philosopher's tendency to mix precision and exposition willy-nilly. Danto worries appropriately over the proper use of the word "is" when it is used to mean artistic identification. Probably 20 percent of the paper concerns the precise definition of this kind of "is." Yet in his first paragraph he refutes Socrates' claim that "art is imitation" with the following argument: "If a mirror image of o is indeed an imitation of o, then, if art is imitation, mirror images are art."[35]

Unless Professor Danto wants to tell us exactly what he means by the use of "is" in this argument he might as well assert: If a cat is indeed an animal, then if a dog is an animal, dogs are cats.

We will not let this silliness detract from the seriousness of the paper and the useful notion of the artworld.

One who has taken Danto's work most seriously is the
philosopher George Dickie. To be sure, much of Dickie's
writing about Danto's artworld ideas has been attempted
refutation. Nevertheless, the theory which Dickie has himself propounded—the so-called Institutional Theory of Art
—owes enough to Danto that it is often referred to as the
Dickie–Danto theory.

What Dickie has done is take the institutional/social
aspects of Danto's artworld and made them the basis for his
theory. Dickie thinks that Danto has not, in his 1964 paper
and in later writings, *established* his contention that art
theories help us identify artworks nor has he *demonstrated*
that art theories make art possible. Moreover, Dickie believes that Danto's careful analysis of "artistically relative
predicates" may be irrelevant to his thesis since Danto does
not establish his claim that having a pair of applicable
predicates constitutes a necessary condition of an object to
become an artwork. Nevertheless, Dickie states flatly "that
Arthur Danto's, 'The Artworld,' inspired the creation of
the institutional theory of art."[36]

Dickie's arguments, as you may suppose, are technical
and we need not consider the details. A good source for
them is Dickie's 1984 book, *The Art Circle.* This book re-
vises the "institutional theory" which first appeared a decade earlier in *Art and the Aesthetic.* (Comments on the
work of both Danto and Dickie can also be found in several
of the articles in the book *Aesthetics and Arts Education,*
mentioned earlier.)

Essentially, Dickie took from Danto the institutional
complex notion of artworld while rejecting the latter's contention that the "theories" of this world identify art and,
indeed, make art possible. In Dickie's theory the artworld is
more of a social and cultural organization while Danto fo-

cuses more on the language through which that world identifies art. This emphasis on language accounts for the analytic difficulty of the last part of Danto's "Artworld" paper
and is what sets the work firmly in the field of analytic
philosophy. Danto says he is out to identify "necessary
conditions" for art. Dickie wants conditions which are
both "necessary and sufficient."

Think of p and q as two English language statements.
To say that "q is a necessary condition for p" means (by
definition) that "p implies q." This latter phrase is called an
implication and is denoted by the symbol $p \rightarrow q$. Precisely,
$p \rightarrow q$ means that q is true whenever p is true. Whenever the
implication $p \rightarrow q$ is valid we say that "p is a sufficient
condition for q."

For example, let p be the statement: "Felix is a cat."
Let q denote the statement: "Felix is an animal." Certainly,
"Felix is a cat" implies "Felix is an animal." So, "Felix is
an animal" is a necessary condition that "Felix is a cat."
Simply put, in order that a thing be a cat, it is *necessary* that
the thing be an animal; in order that a thing be an animal, it
is *sufficient* that the thing be a cat.

On the other hand, the implication $q \rightarrow p$ is *not* valid.
So q is not a sufficient condition for p. (Notice the interchange of the roles of p and q here.) That is, in order to be a
cat it is not sufficient to be an animal.

In general, if $p \rightarrow q$ and $q \rightarrow p$ are both valid (written
$p \leftrightarrow q$) we say that p is a necessary and sufficient condition
for q. Thus, $p \leftrightarrow q$ means both "if p then q" and "if q then
p." (Sometimes we just say "p if and only if q.")

If p and q are mathematical statements rather than nat-
ural language statements and if $p \rightarrow q$ is valid, then $p \rightarrow q$ is
called a *theorem*. Theorems constitute mathematics.

My minor criticism of Danto's handling of Socrates'
imitation argument involves these ideas. When Socrates

says: "Art is imitation," I assume he means, "If x is art then x is imitation." Thus, I take Socrates to say: "Imitation is necessary for art." Evidently, Professor Danto thinks Socrates meant "Imitation is necessary and sufficient for art." (This is not a usual interpretation. If I say: "Helen is fair," you are not likely to think I mean: "Something is fair if and only if it is Helen.")

Philosophers have long searched for necessary and sufficient conditions for art. Obviously, it would be desirable to have a meaningful, intelligible statement q, perhaps a statement prescribing a small set of conditions, such that we have the *valid* double implication: "x is a work of art" if and only if "q holds for x."

Most aesthetic theories, like Danto's "Artworld," seem to yield only necessary conditions for art. In Dickie's language, they are too "thin" to do otherwise. Moreover, Dickie believes many theories are inherently flawed because they too narrowly focus on the art objects themselves. (The imitation theory is a classical example.) One of the appealing features of Danto's "Artworld" is that it recognizes the essential role that the background of artworks plays in determining what is, or is not, art. This "background" notion is what Dickie used as base for his institutional theory. From this, he claims to have produced conditions for art that are both necessary and sufficient.

Here's a nice one-sentence summary from *The Art Circle:* "The institutional theory sets works of art in a complex framework in which an artist in creating art fulfills a historically developed role for a more or less prepared public."[37]

He continues by stating neatly the relationship of the framework to the works of art[38]:

> . . . being a thing of a kind which is presented to an artworld public is a necessary condition for being a work of art. This claim implies another rule of art-making: If one wishes to make

a work of art, one must do so by creating a thing of a kind which is presented to an artworld public. The two rules are jointly sufficient for making works of art. The rules may appear to be satisfied by things which are not works of art, but it must be remembered that the rules operate *within* a specific, historically developed, cultural domain.

Dickie tries to make these ideas clear when he later sets down five "definitions" of the critical terms of the theory. He recognizes (without apology) that the definitions are "circular" but he does not concede this to be a "grave logical error." In fact, he claims that the circularity reveals the *inflected nature of art,* by which he means "a nature the elements of which bend in on, presuppose, and support one another."[39]

This is not analytic philosophy with its axioms, its precise definitions, and its rules of inference. Dickie's definitions have the circularity of Webster's Dictionary but they allow us to see art outside of the center which consists of the art objects themselves. The institutional theory provides a framework within which each of existing art theories—imitation, impressionistic, abstract, etc.—may or may not operate. The framework identifies no specific artistic theory, nor does it rule out any of them. What it provides is something pragmatic, something we can use. And, as we shall see, something we can use for mathematics.

Here, in a nutshell, is the institutional theory. Here are Dickie's definitions from *The Art Circle*[40]:

> I) An artist is a person who participates with understanding in the making of a work of art.
> II) A work of art is an artifact of a kind created to be presented to an artworld public.
> III) A public is a set of persons the members of which are prepared in some degree to understand an object which is presented to them.

IV) The artworld is the totality of all artworld systems.

V) An artworld system is a framework for the presentation of
a work of art by an artist to an artworld public.

You can clearly see the circularity of the definitions.
For example, if you trace the notion of "artist" from I
through IV you see it leads to the concept of "artworld
systems." But when you go to V to determine just what this
may be you find yourself back at "artist" and definition I.

But, given Dickie's purpose, the circularity is not fatal.
It is not, as Dickie put it, a "vicious" circularity and perhaps not even, in this context, a fault. (A more important issue is whether or not the definitions are self-
contradictory. I think not.)

I am certain the circularity makes the analytic philosophers shudder. And it ought for their work includes the task
of making language of this sort precise. Circularity is not
precision.

Paradoxically, it might not trouble the mathematicians—the single set of persons for whom precision is paramount. Mathematics, as we have seen, is precise or else it is
nothing. But Dickie is not talking mathematics. He is not
even talking art but rather is talking about art (just as I am
here talking about philosophy). Mathematicians, I think,
will see little need of precision in these matters. Most of
them will, I believe, doubt that precision here is possible.
They, themselves, talk this way when they talk *about*
mathematics.

Ask a mathematician to define himself and his subject.
Here's what he will tell you: "Mathematics is what mathematicians do. A mathematician is one who does
mathematics."

Circular, sure. But understandable to persons who are,
as Dickie says, "prepared in some degree."

THE MATHWORLD

A fundamental notion in mathematics is the notion of *isomorphism*. Roughly, an isomorphism is a one-to-one correspondence between two sets which preserves particular operations on these sets. When such correspondence exists, the sets are called *isomorphic*. Except for notation, isomorphic sets essentially are identical and mathematicians regard them as the same. We've seen the idea illustrated with different language when we talked about associating each complex number of the form $(x, 0)$ with the real number x. This association becomes an isomorphism once the term is precisely defined. So, mathematicians—and all students of complex analysis—regard the complex number $(x, 0)$ and the real number x as being identical. In this case, it is convenient to think of the reals as being *embedded* in the complex numbers by the isomorphism. (A person who dislikes football might say cynically: "Any two NFL teams are isomorphic." By this he means the two teams have the same number of players and, except for uniform colors, are identical.)

Part of the attractiveness of Dickie's extension of Danto's artworld notion is its intuitive familiarity. We all know that such an institutional complex exists and we have a reasonable notion of the people who move in it: artists, gallery owners, writers, sculptors, critics, photographers, directors, producers, and on and on. Moreover, we understand that some of these players are more central to the institution than are others. We recognize that participation in the artworld is possible on many levels: elementary school art on one end and perhaps Wagner's *Ring* on the other.

Moreover, it is clear that we can come closer to the central activities of this world only if we are willing to invest time and effort in learning to see, to hear, and to appreciate art as the core personnel of the artworld see, hear, and appreciate. For Danto, this requires, in particular, that we master the notion of the *is of artistic identification* and, thus, understand what constitutes a work of art. Dickie does not concern himself with the subtlety of this *is* but insists, rather, that participation in the artworld requires "preparation" and "understanding." Presumably, this comes through a process of formal or informal study of this world; study which might include in the advanced stages a coming to grips with Danto's *is* and the properties of his "artistically relevant predicates."

Danto and then Dickie have attempted to set down the characteristics of the artworld so that we could speak of it other than descriptively and so that the concept has value in a theoretical sense. The attention their ideas have received from contemporary philosophers—both pro and con—show that they are on to something seminal. The Danto–Dickie ideas have touched a collective nerve. They have illuminated the artworld.

There exists another world with properties similar to the artworld. In fact, this second world can be described with a set of five "definitions" which, except for notation, are identical to Dickie's artworld definitions I–V. Therefore, this new world is isomorphic to the artworld in the sense indicated above. Consequently, this second world should be considered identical to the artworld, or more properly, identical to some subset of the artworld.

But there is a significant difference between the two. While the existence of the artworld is intuitively obvious to each educated person, this new world is unknown to nearly all of them. One of its members, Paul Halmos, has written: "It saddens me that educated people do not even know my subject exists."[41]

The subject, of course, is mathematics and Paul Halmos is a mathematician. The unknown world I call the *mathworld*. It contains mathematical objects as artifacts and mathematicians as artists. The present book exists primarily to bring the mathworld out into the light, to instill in otherwise knowledgeable people some understanding of what goes on in it, and to make a first step toward the preparation of a mathworld public.

Here are the five definitions:

1. A mathematician is a person who participates with understanding in the making of a work of mathematics.
2. A work of mathematics is an artifact of a kind created to be presented to a mathworld public.
3. A public is a set of persons the members of which are prepared in some degree to understand an object which is presented to them.
4. The mathworld is the totality of all mathworld systems.
5. A mathworld system is a framework for the presentation of a work of mathematics by a mathematician to a mathworld public.

These definitions are, as you see, identical to Dickie's except there has been a transformation of certain terms. This transformation—indicated by arrows—looks like:

artist → mathematician

art → mathematics

artworld → mathworld

artworld systems → mathworld systems

If we consider certain words in Dickie's definitions as

being artworld "operations", i.e., *participates, making,
work, presented, created,* and *presentation,* we note that
they carry over unchanged to the mathworld. So, except for
the notational changes given by the arrows, the definitions
are the same.

I do not want to worry too much here about the word
"artifact" which appears in each set of definitions. A considerable philosophical literature concerns itself with
which "artifacts" can, or cannot, be works of art. In the
artworld, as Dickie defines it, the notion seems intuitively
clear. We think of an artifact as an art object and, as examples, we think of paintings, sculpture, poetry, and the like.
Mostly, we consider these to be *real* objects, as things made
of canvas and paint, of marble, or as lines of words printed
on a page. But it is easy to go more deeply, as, for example,
does the contemporary philosopher Richard Wollheim, in
his book *Art and Its Objects.*[42] One is quickly led to question
whether a nonreal object can be a work of art. Wollheim
examines the alternate question of whether or not a real
object can be an art object if it is true that art artifacts must
involve an expression of emotion as many philosophers
believe.

We need not pursue these matters far. But we must
acknowledge the significant difference between a mathworld artifact and an artworld artifact. Mathematical objects, as we have seen, are objects of the imagination. They
are abstract notions which have certain properties or satisfy certain conditions which have been imposed on them
at their creation or which have been deduced for them from
more fundamental mathematical principles. They live in
the mathematical, and not the real, world. To be sure, many
of these ideas are abstractions of real objects and have been
specifically conceived to reflect some real object. But they,
themselves, are not real. When Socrates asked the slave boy
Meno about the problem of constructing a square with

twice the area of another, he was not talking about a figure drawn in the sand. The question concerned an abstraction, a mathematical object called a *square*.

Objects of this nature are what mathematicians (as artists) present to the mathworld public. The presentation normally comes through a part of the mathworld system called a "research paper." Consequently, the artifacts are represented by symbols and sentences written on paper as Shakespeare's *Sonnets* are represented on paper and presented to the artworld public. A careful analysis of the differences (or the similarities) between the nature of mathworld objects like equations and artworld artifacts such as poems would take us very far afield. Nothing will be lost at this level if we, in fact, think of mathworld artifacts in terms of their representation as they are presented via papers, books, lectures, or whatever. But we must always understand that the true mathworld artifact lives in the mathematical world: a world of ideas.

We can, incidentally, take comfort in the fact that some weighty philosophers have concluded that imaginary objects can rightly be works of art. One who did was the Oxford philosopher R. G. Collingwood.

Collingwood's book *The Principles of Art* was first published in 1938. Although Collingwood's basic approach to aesthetics is through language rather than perception, *Principles* is not a work of analytic philosophy. It may well be that this book stands as a last attempt to set down a "grand theory of art" before the advent of analytic philosophy which, through its concentration to detail and precise language, made the concept of such theories anachronistic.

Collingwood's philosophical standing can be deduced from the fact that contemporary philosophers like Richard Wollheim still work at refuting his ideas and that today—almost half a century after his death—five of his books are still in print.

The Principles of Art reads easily because Collingwood writes beautifully. But the ideas contained therein are not facile and are not subject to the sort of summaries you hear of books or films as you stand around a cocktail party, drink in hand. Professor Collingwood set himself the goal of nothing less than the description of what constitutes art. To do this, he first considered some common misunderstandings about the notion of art and carefully disposed, one-by-one, of several of the categories which he claims are "of art falsely so called."

Thus, he examines most carefully the common and erroneous confusion of what he calls "art proper" with activities and objects which more properly belong to crafts, to amusements, or to magic. Only when he is finished discussing what art is not does he come to his main task of bringing forward the characteristics of real art. But this evolves most slowly and only after this is complete does Collingwood advance his own "theory of art." Then he writes[43]:

> . . . a work of art in the proper sense of that phrase is not an artifact, not a bodily or perceptible thing fabricated by an artist, but something existing solely in the artist's head, a creature of his imagination; and not only a visual or auditory imagination, but a total imaginative experience. . . . A work of art need not be what we should call a real thing. It may be what we call an imaginary thing. A disturbance, or a nuisance, or a navy, or the like, is not created at all until it is created as a thing having its place in the real world. But a work of art may be completely created when it has been created as a thing whose only place is in the artist's mind. . . . The work of art proper is not something seen or heard but something imagined.

Collingwood did not deal with mathematics in *Principles of Art* and when he described art as "something imagined" it was not complex numbers or differential equations he had in mind. Nevertheless, mathworld artifacts are objects of this kind and it is comforting to note that they can be nicely fitted into Collingwood's aesthetics.

There exist, of course, differences between the math-world and the artworld other than those given by the dissimilar nature of their artifacts. (Our notion of correspondence between these two worlds comes from the nontechnical isomorphism between their defining sentences. The correspondence resulted from the identification of the artworld terms "artist," "art," "artworld," and "artworld systems" with the corresponding mathworld names.) For example, one of these systems incorporates *critics* and *criticism.* The other does not.

We have observed that the institutional theory of art neither incorporates nor rules out any particular artistic theory. Within the artworld framework one can fit the classical imitation theory or some newer theory dealing with, let us say, minimal abstract art. This flexibility of the institutional theory is one of its attributes.

Moreover, the artworld system prescribed in the defining sentences can (and does) allow for the presence of art critics and, hence, for a myriad of theories related to criticism. If you wish, you can within the framework deal with the "deconstructionism" literary theory of Jacques Derrida and Paul de Man or with the "new criticism" associated with earlier theorists like Cleanth Brooks and Robert Penn Warren. Or you may choose the apparent successor to deconstructionism, something called "new historicism." It makes no difference. The artworld framework incorporates either with aplomb.

Theoretically, the same holds for the mathworld. The mathworld framework allows you to criticize mathematics from the point of view of the written text only or from the political and historical atmosphere which surrounded the mathematicians who produced it. Or you might, following Derrida and de Man, attempt to show that the apparent precision of mathematics is illusionary and that the subject is properly interpreted only in the light of some sort of skeptical relativism. You can criticize any way you wish in the sense that the mathworld concept does not say you cannot (it might even be fun although I doubt that two of the mentioned possibilities would lead anywhere). The mathworld *allows* for the possibility of mathematics critics of all types.

But in truth there are none. No critics live in the mathworld system. Although the defining sentences for the mathworld agree exactly, except for certain name changes, with those of the artworld, the associated systems differ significantly. And one of these differences is this: the artworld system includes a plenitude of people whose profession is that of criticism; the mathworld system includes none.

Of course, mathematicians sometimes make judgments or evaluations of another's mathematical work. They most clearly perform these duties when they serve on university promotion committees and are specifically asked for recommendations regarding the quality of a particular tenure candidate's research. Mathematicians also write reviews of research articles and serve as prepublication referees for mathematical journals. But these evaluations normally concern only the "correctness" of the research and its place in the mathematical literature. As they play these criticlike roles, the mathematicians often do address properties of the research such as depth, generality, or significance. But they rarely frame their comments in such a way as to have given even the appearance of what the artworld professionals think of as criticism. G. H. Hardy, I am afraid, spoke for all succeeding mathematicians when he wrote on the first page of his 1940 book, *A Mathematician's Apology:* ". . . there is no scorn more profound, or on the whole more justifiable, than that of the men who make for the men who explain. Exposition, criticism, appreciation, is work for second-rate minds."[44]

Mathematicians make mathematics; they are not critics.

But this is not the main point. Except for the (relatively minor) roles described above there exists no mathematical criticism *except that done by the mathematicians themselves.* The analogy in the artworld would be no art criticism except by artists, no literary criticism other than by writers, and no music criticism except by musicians.

You may, if you wish, argue that such an artworld situation would not be a bad thing. I concede—as do my musician friends—the desirability of removing from newspapers the alleged critical performance reviews thrown together between 11:00 A.M. and midnight by dilettantes who cannot distinguish a violin from a viola at twenty paces. Music criticism should require something more than just access to people who buy ink by the gallon.

On the other hand, the artworld system includes many knowledgeable critics whose presence enriches that world and, in an appropriate and nontrivial way, regulates it. Each artist—painter, writer, musician, or whatever—knows that the presence of external critics requires of him a certain accountability to the artworld public and not just to colleagues within his own art circle. He may resent this accountability, but he ignores it at his peril. For the artworld public, by and large, pays attention to the critics and you can measure the level of participation of a member of the public by the seriousness of the criticism he reads. The more knowledgeable the member, the more serious the criticism to which he turns. Moreover, the very existence of an artworld public, at least at the lower levels of participation, may well depend on the presence of critics who, through public media, keep art in view.

No analogy holds for mathematics. The mathworld

contains no critics. Its specialized artists, the mathematicians, are accountable only to one another. There are no mathworld critics to keep mathematics in public view. And, partially because of this, there is scarcely a mathworld public at all.

It is interesting to note that the third definition in Dickie's list is identical, word for word, to definition #3 in the mathworld list. A public is a public so long as it "understands." But while the artworld public is huge, the mathworld public is tiny. Outside of the mathematicians themselves as creating artists, it consists mainly of those relatively few persons with true mathematical understanding: mainly some of those who *use* mathematics and some of those who *teach* mathematics. But not all, even, of these. You can use mathematics and not appreciate it and—what is worse—you may be allowed to teach mathematics without understanding it at all. And, when you teach mathematics without understanding, you are unlikely to contribute to increasing the size of the mathworld public.

What needs to be done seems clear. There is no question of the similarity between the mathworld and the artworld. Mathematicians act and think like artists as they create mathematics. But, at this moment, the mathworld stands apart from the artworld. They are *disjoint* worlds. The two do not intersect even though there exists an isomorphism from one to the other. What we must do is somehow *embed* the mathworld in the artworld. Alas, this cannot be done on paper.

A true embedding—a real-world identification of the mathworld as a subset of the artworld—will require changes affecting three groups of people. Namely,

    i. an increased willingness of mathematicians to talk *about* mathematics,

   ii. a beneficial change in the way young people are
        taught mathematics,
  iii. a positive change in attitudes of nonmathemati-
        cally educated adults toward mathematics.

I've had no luck with item i. As we have seen, research mathematicians by and large do not want to talk about mathematics. Most research mathematicians work in universities where they are paid to do mathematics, not to talk about it. Exposition, no matter how valuable or how well-done, falls outside the university reward system. Item ii will require nothing short of a revolution in mathematics teaching. And, if I am correct in my assessment of this activity, the new teaching must include a significant component of "aesthetical mathematics" along the lines of Poincaré and Papert. (We are not, incidentally, headed that way. The future seems to hold, for mathematics instruction, an increased dependence on technology in the form of computers, hand-held calculators, and video presentation. Surely, I cannot be the only person to notice the clear correlation between the declining mathematical abilities of American students and the aggressive introduction of these technologies into the mathematics classroom. The decline began as the technology came in.)

The change I call for in item iii will one day come about. There will come a time when mathematical ignorance, like public smoking, will become socially unacceptable. Then attitudes will change. In the meantime, item iii is a primary purpose of this book.

TRUTH AND BEAUTY

When John Keats asserted "Beauty is truth, truth beauty," he was not thinking of mathematics. Instead, he had in mind something almost opposite. As you know, Keats and his Romantic contemporaries, Byron and Shelley, mistrusted the Enlightenment's attempt to reach truth by consecutive reasoning. Keats felt that one could reach "the still heart of life," not by means of any intellectual pattern but rather through some sort of poetic insight to nature's inscrutable mysteries. He once wrote to his brother, George: "A man's life is not subject—happily—to mathematical analysis."

But Keats did not know mathematics and, consequently, could draw no connection between it and beauty as he did between truth and beauty. Keats, however, knew poetry and great poetry often says more than the poet intended, more even than the poet believed.

For mathematics stands as art in the mathworld as poetry stands as art in the artworld. And in the mathematical world of symbols and equations and inferences lives the sharpest kind of truth. "Beauty is truth, truth beauty" or "mathematics is truth, truth mathematics." Take your choice. I like to think they are the same, that Mr. Keats said more than he meant.

I suspect Keats had in mind, perhaps unconsciously, something similar to Mortimer Adler's notion of two kinds of beauty: *enjoyable beauty* and *admirable beauty*. These notions appear in Adler's book *Six Great Ideas* and are used to divide the collection of "beautiful objects" into two classes. Objects of the first class are those which bring pleasure upon being contemplated. These objects—those having *enjoyable beauty*—depend, along with the pleasure they bring, on the "taste" of the observer. To most people, for example, mathematics does not have enjoyable beauty.

Objects with *admirable beauty* are those which have been judged by experts as being beautiful—where, by "experts," Adler means individuals learned enough or experienced enough to have developed superior taste. Mathematicians have superior taste in mathematics. Were the subject properly taught, so would many others.

These notions—as described in detail by Adler—seem reasonable and commonsensical. The Julian Schnabel paintings I saw in New York in early 1990, for example, did not bring to me enjoyable beauty. Yet, I may well concede their admirable beauty once I've learned more about them.

Moreover, the admirable beauty notion meshes nicely with the artworld concept and, hence, with the mathworld idea. One of the defining characteristics of the artworld requires a degree of "understanding" of the artworld public. In Adler's terms, this requires a public which recognizes the notion of admirable beauty.

Mr. Adler also agrees that admirable beauty—because it is associated with patient study and a steady perfection of taste—represents a special kind of truth. Admirable beauty is beauty for experts, for those equipped to know the truth.

John Keats, I suspect, considered himself expert in matters having to do with beauty. If so, his idea of beauty may be close to Adler's notion of admirable beauty. If the two ideas are identical then Keats may have been saying (as poets often do) more than he intended. If you dare go one step further and slip mathematics into the collection of admirably beautiful objects then you may have Keats's endorsement of the mathworld concept.

I do not find this strange. Much of mathematics is concerned with the identification of one thing with another, like the isomorphic relationship between the real numbers and the subcollection of complex numbers of the form $(x, 0)$. It is easy for mathematicians to slide this concept over into the real world. It is even easier for poets.

In a recent *New York Times* review, Doug Anderson wrote the following lines about an essay of Howard Nemerov[45]:

My favorite essay, "On Metaphor," begins with an anecdote about the author trying to identify with certainty some purple finches that landed on his lawn. He takes down Peterson's "Field Guide" and, after reading an exacting, if prosaic, scientific description of the bird, comes to the line "a sparrow dipped in raspberry juice." He concludes: "Now I *know* that I am seeing purple finches."

Keats would understand. So do I.

PRINCIPLES

George Dickie's institutional theory provides a framework through which the artworld determines what is, or is not, art. But the framework does not distinguish good art from bad. For this, something else is required: an artistic theory, perhaps. Or, maybe just plain good taste. The situation in the mathworld is the same. Again, the institutional framework determines what constitutes mathematics. *Good mathematics* is another matter. Just as in the artworld, something else is needed.

We need to be careful here. We are speaking of mathematics as art because of the isomorphism between the artworld and the mathworld given by their defining equations. So, good mathematics here means mathematics as good art. That is, by a *good mathematical artifact* we will mean a *beautiful mathematical artifact.* Or, as the mathematicians say, we mean *elegant mathematics.* How does one distinguish the elegant from the ugly? Mathematicians seem to know instinctively one from the other. How do they tell?

It is hard to overestimate the degree of unanimity which exists among mathematicians regarding the beauty of certain mathematical results. This unanimity of agreement far exceeds that which I have encountered in other areas of arts. You will have no trouble finding, for example, a collection of string players with varying opinions regarding the beauty of the Brahms Violin Concerto. It seems, in fact, almost a matter of honor for certain violinists to dismiss this piece—which to many of us stands among the wonders of the world—as overly sentimental and romantic. Moreover, once, in my presence, two contemporary painters carried forth a serious discussion on whether or not the paintings of Andrew Wyeth represented real art or merely some combination of craft and amusement.

But mathematicians, by and large, do not disagree on which mathematics is elegant and which is not. You cannot find, for example, a research mathematician who will argue that the Pythagorean proof of the irrationality of the square root of 2 does not have aesthetic value. I, in fact, cannot conceive that such a mathematician could possibly exist. Indeed, if Poincaré is correct about the essence of creative mathematics being bound up with the concept of aesthetic experience, such a person could not exist. A person who fails to see the beauty in this theorem and its proof could not be a mathematician in the first place.

Consequently, the mathematicians must work from some set of commonly accepted aesthetic principles. Otherwise, how could they so completely agree on the beauty—or the lack of beauty—of a particular theorem. Yet, there exists no aesthetic theory of mathematics to which they are signatories. Somewhere then, there must reside at least a handful of practical principles which they share and which leads them to collectively agree on a particular mathematical result and to nod knowingly and simultaneously and to say together and in chorus: "Isn't it elegant?"

You can find, in one place or another, vague attempts to write down such principles. G. H. Hardy, you will remember, when he tried to come to terms with what made certain mathematical results elegant—such as the proof of the irrationality of the square root of 2—brought out a packet of adjectives such as *seriousness, depth, economy,* and *generality.* And when he applied these adjectives to the description of certain theorems, he just skirted the formulation of the kind of principles I have in mind. Please understand I do not mean to imply that Hardy was being vague. You would have to look hard to find anyone who practiced a higher degree of precision. I am just suggesting that he came only close to writing down practical principles which might help outsiders understand what mathematicians consider when they speak to one another of the beauty of mathematics. I will try to be more specific.

I have identified two principles which—I believe—are regarded by mathematicians as standards by which the aesthetic quality of a mathematical notion can be gauged. The first is the *principle of minimal completeness* and the second I call *the principle of maximal applicability:*

1. *A mathematical notion* N *satisfies the principle of minimal completeness provided that* N *contains within itself all properties necessary to fulfill its mathematical mission, but* N *contains no extraneous properties.*
2. *A mathematical notion* N *satisfies the principle of maximal applicability provided that* N *contains properties which are widely applicable to mathematical notions other than* N.

(Occasionally, I will refer to these as simply the "minimal principle" and the "maximal principle," respectively).

The word "notion" has here been deliberately chosen. More often than not, mathematicians apply the terms "beauty" and "elegance" to a theorem and/or to the proof of the theorem. But, they may also apply these terms to other mathematical objects such as an equation, an inequality, or even a definition. So, in the principles, I allow the idea of "notion" to include any of these. Indeed, as far as the principles are concerned, N may be any object which lives in the mathematical world of Figure 1.

The principle of minimal completeness is reminiscent of the philosophical principle known as "Occam's razor" which says: "Entities are not to be multiplied without necessity." However, Occam's razor usually applies to propositions of the form "p implies q" and represents an admonishment that the hypothesis p should be kept as "thin" as possible while still allowing the conclusion q to be drawn. I mean something more general by the first principle given above. I mean that it applies to all mathematical objects and not just to statements of the form "p implies q."

In terms of the first principle, a mathematical notion has aesthetic value provided it is, in some sense, complete or self-contained but yet contains nothing extraneous. In the second principle, the word "applicability" does not refer to the notion's utility in explaining or describing real-world phenomena but rather to the relevance of N to mathematics itself. This principle demands that the notion not be just an isolated proposition of only local interest—no matter how nice it may seem at first glance. If, for example, you show a fourth-grade student our earlier "method" for dividing 64 by 16 by simply writing the appropriate fraction and canceling the 6s,

$$\frac{\cancel{6}4}{1\cancel{6}} = 4,$$

he will likely proclaim the result to be "neat." Yet it has no mathematical value—aesthetic or otherwise—because the method has no applicability beyond itself. Try it, in fact, on almost any other quotient of two-digit numbers and you'll get a wrong answer.

On the other hand consider the solution to the following simple counting problem which I learned from the mathematician Paul Halmos. Consider a single-elimination basketball tournament which begins with 8 teams. There is no difficulty in determining the number of games necessary to produce a winner. There will be four games in the first round, two in the second round, and then one more game in the tournament's "finals." Thus, a total of $4 + 2 + 1 = 7$ games will be played. However, this method does not generalize to an *arbitrary* number of teams. Consequently, this crude counting method fails completely to satisfy the principle of maximum applicability. Suppose our tournament begins with, say, 1,729 teams. In this case, lots are drawn and the odd team sits out the first round. The winning teams advance and the procedure is repeated until each team but one has lost a game. The remaining team wins the tournament. Now, how many games are played?

Here's the Halmos method of solution. Notice that each game has exactly one loser and that each team, except the tournament champion, will lose exactly one game. Thus, there are exactly as many games as there are losing teams. And if we begin with 1,729 teams there will be 1,728 losers. So, exactly 1,728 games are played.

Halmos says this "pure thought" method of solution is "pretty." An understatement, Mr. Halmos. It's elegant.

Let's look a bit more closely at the Halmos solution of this simple counting problem. Why is it elegant? In the first place, the solution is completely self-contained. It relies on nothing extraneous or preliminary except the rudimentary understanding of the phrase "single elimination" and the commonplace knowledge that a basketball game never ends in a tie. And it would be difficult to conceive an alternate yet more reduced solution to the problem. The Halmos solution can, in fact, be written in nine words: The number of games equals the number of losers. Consequently, the solution satisfies the principle of minimal completeness—it says what must be said, and no more.

Moreover, the solution depends on a notion which has great applicability throughout mathematics—the concept of one-to-one correspondence between sets of objects. The problem asked us to count the number of games played in a single-elimination basketball tournament which began with 1,729 teams. If you consider the problem directly, the counting seems difficult. It is hard to get a handle on the manner in which the games will be played because of the relatively large number of teams with which we start and the fact that lots must be drawn after each round to determine which team does not play in the next round. Counting the games directly is not easy.

On the other hand, counting the number of losers is a triviality. There will be one champion and 1,728 losers. The idea which leads us to what Halmos calls "the inspired solution" is the observation that there are exactly as many games as losers, that is, that the set of losing teams and the set of games played may be placed in one-to-one correspondence with each other.

The notion of one-to-one correspondence consists of the following. Consider any two sets of objects. Call the sets A and B. The objects contained in A and B may be of any type whatever—people in a certain room, trees in a particular backyard, or gold watches in Cartier's window. In our counting problem, A consisted of the set of games and B consisted of the set of losing teams. In mathematics, A and B are often sets of numbers of one kind or another. If, for each element a belonging to A there exists a unique element b in the set B and, conversely, if for each b in B there exists a unique member a belonging to A, we say that there exists a one-to-one correspondence between the sets A and B. In other words, A and B are in one-to-one correspondence with each other if the members of the two sets can be paired with one another in such a way that the pairing process uses each element of *A* and each element of *B* and uses these elements only once.

What you do, of course, when you *count* the number of elements in a set is to place the set in a one-to-one correspondence with a subset of the natural numbers, 1, 2, 3, ... To say, for example, that there are three pine trees in your backyard is to say that the set of these trees can be placed in a one-to-one correspondence with the set containing the numbers 1, 2, 3. You can make this correspondence in many ways. One way would be to step outside with a brush and bucket and paint the number 1 on one pine tree, the number 2 on another, and the number 3 on the remaining tree. Another way—the one most commonly used—is to stand at your window, point to the trees one by one and count 1, 2, 3 as your finger moves your line of sight from one tree to another.

Inspired by Paul Halmos, we have used the method of one-to-one correspondence to solve an admittedly commonplace and uninteresting problem, that of counting the number of games played in a certain basketball tournament. But, although the problem is commonplace, the method is not. It, in fact, permeates all of mathematics and has important applicability to algebra, analysis, topology, and all other subareas of the subject. It was George Cantor's (1845–1918) great idea to apply the notion of one-to-one correspondence to infinite sets which allowed mathematicians for the first time to come to exact grips with the concept of *size* of infinities.

Cantor's work was truly revolutionary and ultimately brought to him the kind of mathematical immortality that caused David Hilbert (one candidate for this century's greatest mathematician award) to say "No one shall expel us from the paradise which Cantor has created for us."[46]

Like most intellectual revolutionaries, Cantor was con-

siderably abused before he was canonized. Leopold Kronecker (1823–1891), for example, described Cantor's work
as "humbug" and tried to prevent its publication. And our
own Jules Henri Poincaré said of it, "Later generations will
see Cantor's set theory as a disease from which one has
recovered."[47]

Cantor's reaction to these, and similar attacks on his
work, was to fall into a series of mental breakdowns.

But his work prevailed and mathematics was never
again quite the same. And all—or nearly all—of the ancient
bugaboos regarding sets which had troubled philosophers
from the days of Zeno were finally and simply resolved. For
Cantor showed us that we must deal with infinite sets as an
illiterate man would deal with two leather bags of colored
beads. The man wants to know which bag, the one on the
left containing white beads or the one on the right holding
red beads, contains the greatest number of beads. But our
man is completely illiterate. Not only does he fail to read or
write, he cannot count. Yet, the problem of the greater number of beads presents him no difficulty.

The man simply takes simultaneously a bead from
each bag and places them side by side on the ground. He
then repeats the process until at least one of the bags is
empty of beads. If he runs out of white beads first he concludes that there are more red beads; if the red bead leather
bag empties first, he is certain to have more white beads. If
the bags simultaneously become empty, he concludes that
each bag contained the same number of beads. For, in this
case, what he has done by drawing pairs of beads—one
from each bag—is physically to establish a one-to-one
correspondence between the set of red beads and the set of
white beads. Nothing to it.

Cantor did exactly the same thing with infinite sets.
There are infinitely many positive integers 1, 2, 3, . . . and

there are infinitely many even integers 2, 4, 6, ... But surely it is absurd to believe there are as many even integers as there are positive integers, because the even integers are *contained* in the positive integers. No, said Mr. Cantor. There are *exactly* as many even integers as positive integers. And the way to see this is to pair the elements in the two sets exactly as our illiterate man did with the red and white beads. You can, for example, pair 1 with 2, 2 with 4, 3 with 6, $\cdots$ (in general, you pair n with $2n$). This pairing produces a one-to-one correspondence between the set of positive integers and the set of even integers. Consequently, these two infinite sets are the same size.

Incidentally, this process fails if you try to establish a pairing between the integers and the set of real numbers. One can prove, as Cantor did, that any such attempt will cause the exhaustion of the integers and still leave some real numbers unassigned. Thus, the infinity of the "number" of real numbers exceeds the infinity of the number of integers. In fact, Cantor's simple proof shows that there are more real numbers between 0 and 1 than there are members of the entire set of natural numbers: interesting, surprising, important, and widely applicable.

Consequently, the Halmos solution of the tournament game counting problem uses a method of wide applicability in mathematics. So wide, in fact, I am ready to pronounce that the solution satisfies the principle of maximum applicability as I have earlier argued it satisfied the principle of minimum completeness. Therefore, I proclaim the solution to be "elegant."

I cannot fail to mention one more aspect of the counting problem. The great Indian mathematician Ramanujan came to England in 1914. He was elected Fellow of the Royal Society when he was only thirty years of age. One year later, he was also chosen Fellow of Trinity College and became the first Indian to receive either of these honors. But shortly afterward he fell ill. His friend and colleague G. H. Hardy came to visit. In the preface to Hardy's *Apology*, C. P. Snow describes the following incident[48]:

> Hardy used to visit him, as he lay dying in hospital at Putney. It was during one of those visits that there happened the incident of the taxi-cab number. Hardy had gone out to Putney by taxi, as usual his chosen method of conveyance. He went into the room where Ramanujan was lying. Hardy, always inept about introducing a conversation, said, probably without a greeting, and certainly as his first remark: "I thought the number of my taxi-cab was 1729. It seemed to me rather a dull number." To which Ramanujan replied: "No, Hardy! No, Hardy! It is a very interesting number. It is the smallest number expressible as the sum of two cubes in two different ways."

Yes. One way to express 1,729 as the sum of two cubes is

$$1,729 = (12)^3 + (1)^3.$$

Another way is

$$1,729 = (10)^3 + 9^3$$

Ramanujan says that 1,729 is the smallest integer that has two representations of this type.

Euclid's proof of the infinitude of the prime numbers (p. 83) clearly satisfies both the minimal principle and the maximal principle. (The primes, in fact, *pervade* mathematics and there is no part of the mathematical world where these numbers, or their generalizations, do not have applicability. As for minimality, the proof can essentially be reduced to the single sentence: "If $p_1, p_2, \ldots, p_n$ are *all* the primes then one of them must have the impossible task of dividing the number $p_1 p_2 p_3 \ldots p_n + 1$.)

Similarly, the Pythagorean proof of the irrationality of $\sqrt{2}$ (p. 136) satisfies both principles. (The proof is completely self-contained and can be reduced to not more than 10 lines. So the minimal principle holds. The maximality criterion is satisfied once one notices that the method of proof extends to show that $\sqrt{3}$, $\sqrt{5}$, $\sqrt{7}$, . . . are also irrational.)

Hardy was justified in describing these two theorems as "gems." Since both principles hold, I'll call them "elegant."

On the other hand, the famous "prime number theorem" (p. 84) satisfies one of the principles but not the other. Recall that this theorem gives the rate at which the number of primes less than a given positive integer n tend to infinity as n tends to infinity. The theorem and its proof are central to many areas of mathematics. A key element in the 1896 proof involves the behavior of a certain complex function known as the Riemann zeta function (p. 82). This particular function is not completely understood today, almost a century after it played its role in the proof of the prime number theorem. The location of the "zeros" of the zeta function, i.e., the specific complex numbers on which this function takes the value 0, remains unknown. It is believed that all the "nontrivial" zeros live on a particular vertical line in the complex plane. The precise statement of this conjecture is called the Riemann hypothesis and stands as the paramount unsolved problem in all of mathematics. Any mathematician will confirm my claim that the prime number theorem satisfies the maximal principle.

On the other hand, it does *not* satisfy the principle of minimal completeness. The standard proof requires deep results from the theory of complex analysis and fills many pages of printed text. One could make it self-contained only by extending these many pages to something on the order of a complete book. I find nothing minimal about the proof of this theorem.

Consequently—while I will allow the adjectives "significant," "deep," "fundamental," and "remarkable"—I will not call the prime number theorem "elegant." Because, for this theorem, the minimal principle fails.

It's not hard to give examples of the other way around, i.e., that satisfy the minimal principle but not the maximal principle. Consider the statement: "The sum of any positive number and its reciprocal is at least 2." Precisely, this translates to the elementary

Theorem: If $x > 0$, then $x + \dfrac{1}{x} \geq 2$.

Here's a *proof.*

$$(x - 1)^2 \geq 0.$$

Thus,

$$x^2 - 2x + 1 \geq 0,$$

so

$$x^2 + 1 \geq 2x,$$

and hence,

$$x + \frac{1}{x} \geq 2.$$

The proof could hardly be shorter. Moreover, it is self-contained in the sense that nothing is required other than the facts that the square of any real number is nonnegative and that the "sense" of an inequality is preserved by addition of equal quantities to both sides and by division by positive numbers. So, the principle of minimal completeness holds.

But the principle of maximal applicability certainly does not apply. The result itself is very specific and the proof begins with a "trick." I just pulled the first inequality out of the air the way a magician pulls a rabbit out of a hat. It's a neat proof but hardly an elegant one.

All these examples, so far, have been theorems. But, as we have seen, the N in the minimal principle and the maximal principle can be any kind of mathematical object. We earlier examined (p. 86) the equation $e^{i\pi} + 1 = 0$ from the point of view of the principle of minimal completeness and we noticed that the principle held. One need only make a cursory examination of complex analysis textbooks (to see the wide applicability of the relationship given by the equation) to note that the maximal principle is also satisfied. I know of no equation half as elegant.

Incidentally, I am aware that any equation—if indeed it is correct—can be phrased as a theorem. But, in the case of $e^{i\pi} + 1 = 0$, it is the *equation* and not the corresponding theorem which is elegant. (The corresponding theorem— when stated—would be equivalent to the equation $e^{i\pi} = -1$ which is quite pretty but is not complete because it omits 0.)

One can find elegance in mathematical physics. A clear example is Newton's famous Law of Gravitation. This assertion (often called the "inverse square law") states that the gravitational force F existing between any two bodies in the universe having masses M_1 and M_2, respectively, is given by

$$F = \frac{GM_1M_2}{r^2}.$$

Here r denotes the distance between the two bodies and G is a constant (called the "gravitational constant").

The validity of our two principles in this case is clear and well known. (A more minimal description of the gravitational force between *any* two bodies can hardly be conceived. Physics textbooks are filled with applications of this law—one of which is the deduction from it of the elliptical laws of planetary motion, which were only posited by Kepler.)

As a final example, let's consider a real-world result which we can (surprisingly) establish by pure thought. Let's *prove*—without leaving the room—that there are at least two trees in the world having the same number of leaves. And let's do it in such a way that the result is elegant.

A theorem, you will recall, is a proposition of the form "p implies q" whose validity has been established. We need to formulate our theorem as a proposition of this type and then produce a proof that the proposition holds. So, we must establish a hypothesis p and a conclusion q, and we must write them in proper mathematical form. The hypothesis is what we *assume,* the conclusion is what we *deduce.* We want the hypothesis to be as minimal as possible. Whatever we may mean by "trees" and "leaves," it seems obvious that there are more trees in the world than there are leaves on any single tree (probably many more). I would suspect, in fact, that there are more trees in heavily wooded western Pennsylvania than there are leaves on any single tree in the whole world. So, for our hypothesis, we will assume that the number of trees exceeds the number of leaves on any single tree by some positive number. It is sufficient to assume that the excess is only 1. Our theorem then takes the form:

Theorem: Let t denote the number of trees in the world and let m denote the maximum number of leaves on any single tree. If t exceeds $m + 1$, then there exist at least two trees with the same number of leaves.

Proof: Since m denotes the maximum number of leaves on any one tree, each tree will possess either 0, 1, 2, 3, . . . or m leaves. Imagine $m + 1$ boxes sitting in a row on your floor, each tagged, in order, with the numbers 0, 1, 2, . . . , m. Now imagine bringing the world's trees (appropriately reduced in size) one by one into your room. Place each tree in the box which bears the label equal to that tree's number of leaves (thus, a bare tree goes into box number 0, a tree with one leaf goes into box number 1, a tree with two leaves goes into box number 2, and so on).

We have only $m + 1$ boxes and, by hypothesis, we have more than this number of trees. Hence, some box must contain at least two trees when trees have been brought into the room. Thus, at least two trees must have the same number of leaves (if two trees are in, say, box number 1729, then those two trees have exactly 1729 leaves).

Q.E.D.

In order to see that the proof satisfies the two aesthetic principles, let's notice that the key idea in the proof is a result known in mathematics as the pigeonhole principle. This intuitively obvious principle asserts that, if you have more pigeons than holes, then some hole must contain at least two pigeons. This principle has wide use and great value in the branch of mathematics known as "combinatorics" and, indeed, in all other branches which deal with difficult, finite counting problems. Because of this, we are allowed to claim that the proof satisfies the principle of maximum applicability.

Moreover, if the extraneous parenthetical remarks are removed, the proof becomes appropriately minimal. (A mathematician might write the proof in one line as: "The theorem follows from the pigeonhole principle.") Thus, both the maximal and the minimal principles hold.

Accordingly, the result is elegant. (It is also surprising. We have seemingly established a result about trees and leaves without going outside our room. In fact, we need know nothing whatever about trees or leaves other than $t > m + 1$ to prove the theorem. No observation required, just pure thought.)

AESTHETIC DISTANCE

Each creative mathematician recognizes instinctively the *aesthetic experience* of mathematics. The experience is commonplace; its explanation is rare. But there does exist published aesthetics research to which we can turn for help. One paper, in particular, provides a convenient metaphorical device through which we can discuss the obviously different categories into which various groups of nonmathematicians fall with regard to their appreciation of mathematics.

The two groups which in this regard are most clearly defined are the humanists on the one hand and the scientists and engineers on the other. (We will examine these groups in detail in a subsequent chapter. There, we will analyze more carefully the exact nature of their differing attitudes toward mathematics.) For example, if you select at random a professor of English literature and a professor of industrial engineering, you will find they have in common the fact that neither remotely appreciates mathematics as does a practicing mathematician. But, you will also find that each perceives mathematics in a manner vastly different from the other. The industrial engineer can in no way be considered a mathematician. Yet, he uses the subject—at a more or less elementary level—each day as he goes about the practice of his profession. To the engineer, mathematics represents a necessary tool without which he cannot do his work. But it represents only a tool and the subject brings him no more comfort than a screwdriver brings a carpenter. In some sense, the engineer stands too close to mathematics to see in it anything but a reminder that he has leftover data which must be studied, graphs which must be drawn, and computer printouts which must be read. To the engineer, mathematics represents unfinished work. In terms of degree, the subject is too *hot* to bring him aesthetic pleasure.

The humanist, on the other hand, thinks of mathematics almost not at all. He, like the rest of us, once engaged in a certain amount of compulsory mathematics education. But the English professor distanced himself from the subject as much as possible as soon as he could. He is an English professor and not an engineer exactly because developed within him is a feeling for language and literature and a sense of beauty which seemed to be the exact opposite of the drill and drudgery he faced each day in his mathematics class. Once the required courses were past and he could choose for himself books to read and disciplines to study, he chose them as far away from mathematics as possible.

The humanist, by choice, stands as far from mathematics as he can. And he stands upwind.

In 1912, Cambridge's Edward Bullough[49] published a lengthy paper in the British Journal of Psychology: "Psychical Distance as a Factor in Art and an Aesthetic Principle." In the paper, he set down, in a reasonably complete manner, the concept of distance to which I have alluded in describing the relative positions of the English professor and the industrial engineer with respect to the aesthetics of mathematics: the engineer failing to appreciate the subject because he is too close, the English professor failing because he stands too far away. Bullough tried—through example and argument—to come to grips with this concept in such a manner that it would have value as both a metaphor for the presence or absence of aesthetic pleasure when one is in the presence of a work of art, and as a serious aesthetic principle which could be used to anticipate and to explain this experience.

Evidently, he succeeded. Donald Sherburne, for example, says, "Edward Bullough's theory of the Psychical Distance has become a classic doctrine of aesthetic theory that must be taken into account by all aesthetic thinking."[50] And James L. Jarrett writes of Bullough's ideas, "Perhaps no more influential idea has been introduced into modern aesthetics than that of psychical distance."[51]

I concur that Bullough's theory "must be taken into account" but I see it more as metaphor, more as a means for classifying one's reaction to art rather than as a "theory" which explains the notion of the aesthetic experience. Nevertheless, it has clear value—particularly with respect to mathematics and the presence or absence of an associated sense of pleasure and beauty. In particular, the concept will allow us later to describe, in some systematic manner, the differing attitudes toward mathematics possessed by the members of C. P. Snow's "two cultures."

In the following, I will attempt to describe briefly Bullough's ideas as I interpret them. However, I have taken the liberty to change some of his terminology and to refer to certain geometrical notions of discs and rings. And, to emphasize that what I am saying here is my own version of Bullough's ideas, I will refer to *aesthetic distance* rather than "psychical distance"—a term which has been used by others although perhaps not in the exact sense as do I.

Let's begin by considering an art object and an observer. The object may be a painting or a piece of sculpture. Or it may be something less compact such as an epic poem, a play, or a ballet. The object need only be a perceptible thing somehow brought within the sensory range of the observer and which can be considered by him, consciously or unconsciously, in aesthetic terms. The object might be a violin concerto to which the observer listens by radio transmission or a work of architecture which he views from a hovering helicopter. The object may take many forms. The observer may be brought to it in many different ways. As we have seen, the object may be a mathematical idea. In this case, the observer may be a classroom student.

It will be helpful to denote the art object and the person who observes it by two points in the plane. This has been done in Figure 11, where the object is denoted by the letter A (for art) and the observer by the letter P (for person). As you look at Figure 11, you should think of the relative positions of A and P as being described only in a metaphorical sense. That is, the actual distance between the two points in the figure, as measured, say, by a ruler, bears no relation to the true real-world distance between the observer and the object. Thus, Figure 11 does not represent the real-life situa-

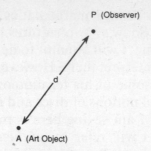

FIGURE 11. Aesthetic distance.

tion of a person standing before Michelangelo's David in the hall of the Accademia in Florence. For, in a sketch of such a situation, the measured distance in Figure 11 would represent—according to some scale—the true distance measured in feet between the person and the base of the great marble statue. Instead, we mean to describe something Bullough describes as "distance in its general connotation." This general connotation Bullough called "psychical distance." In my version of Bullough's ideas, I call it "aesthetic distance." ("Psychical" no longer possesses the philosophical connotation it had in Bullough's day. I prefer the phrase "aesthetic distance" as a reminder that the emphasis here is on philosophical, rather than psychological, matters.) The idea is as follows.

Let us agree to accept the existence of the concept of "aesthetic experience" and let us also agree that—although we may be unable to describe it precisely—we know this experience when we feel it. Thus, we agree that—in the presence of certain art objects—we recognize this aesthetic experience and, in some sense or another, the experience brings us pleasure. Moreover, in order to keep the analysis simple enough to control, we will dismiss the notion of "degrees of aesthetic pleasure" and understand by "aesthetic experience" something which a particular art object brings or does not bring to a particular observer. Clearly, these assumptions are overly simplifying but they are necessary in order that we can make sense of Bullough's basic and far-reaching idea.

Thus, we are asserting that a particular observer either does or does not *have an aesthetic experience* when he contemplates a given art object. In Figure 11, the straight line distance between P and A is now to be thought of as the "aesthetic distance" between the object and the observer. Bullough's concept is that the observer receives from the object an aesthetic experience if and only if this distance remains between certain bounds.

Bullough's idea has been sketched in Figure 12. Here, you see drawn two concentric circles each having object A as its center. The radii of the circles are to be interpreted in terms of the notion of aesthetic distance. The shaded region which lies within the larger circle of radius r_2 and outside the smaller circle of radius r_1 we will call the *aesthetic ring* determined by art object A. According to Bullough's

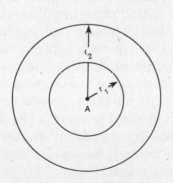

FIGURE 12. The aesthetic ring.

scheme, an observer participates in an aesthetic experience if and only if—in the sense of aesthetic distance—he stands within the aesthetic ring. As Figure 12 is drawn, our observer P stands outside the ring and, consequently, receives no aesthetic pleasure from object A. Bullough would say that the observer is "psychically overdistanced" from the object. If the observer P were to move toward the object— again in the sense of aesthetic distance—so that he came inside the circle of radius r_1, Bullough would say he has become "psychically underdistanced" (my corresponding phrases are "aesthetically overdistanced" and "aesthetically underdistanced," respectively). The object provides the observer aesthetic pleasure only when his aesthetic distance from the object is at least r_1 and not more than r_2, that is, only when he stands within the aesthetic ring.

Overdistancing or underdistancing each produces aesthetic failure. *But the failures are of different kinds.* According to Bullough, an observer who is too close to a work of art will be too practical or subjective toward it to appreciate it properly. If the observer is overdistanced, he will see the object as something cold and withdrawn and he will again fail to appreciate it as art.

A simple but illustrative example of understanding is that of the person Bullough describes as "the proverbial unsophisticated yokel whose chivalrous interference in the play on behalf of the helpless heroine can only be prevented by impressing upon him that they are only pretending."[52]

The yokel, because of his inexperience and lack of training, cannot separate the action on the stage from reality and he rushes forward to remove the damsel—by force of arms if need be—from her distress. Although the yokel's motives might be noble, he is—as Bullough says—"not the ideal type of theatrical audience." He stands—in an aesthetic sense—way too close to the play to take from it any pleasure.

A second, and often quoted, example which Bullough also provides involves the theatre. The example is used by Bullough for two purposes: to illustrate the concept of underdistancing and to show that preparation and predisposition for a work of art will not always lead to heightened aesthetic pleasure. To be sure, Bullough says, preparation increases the *chance* of aesthetic pleasure but the underlying principle—called the *principle of concordance*—requires qualification[53]:

> Suppose a man, who believes he has cause to be jealous of his wife, witnesses a performance of "Othello." He will the more perfectly appreciate the situation, conduct, and character of Othello, the more exactly the feelings and experiences of Othello coincide with his own—at least he ought to on the above principle of concordance. In point of fact, he will probably do anything but appreciate the play. In reality, the concordance will merely render him acutely conscious of his own jealousy; by a sudden reversal of perspective he will no longer see Othello apparently betrayed by Desdemona, but himself in an analogous situation with his own wife. This reversal of perspective is the consequence of the loss of Distance.

Othello yells, "down strumpet," and he places his hands around Desdemona's neck. When he does, the yokel leaps from his seat, races on stage, and saves her by wrestling to the stage floor the bewildered actor playing the role of the Moor of Venice.

Our other theatregoer, the suspicious husband, suffers less dramatically, but he suffers nevertheless. He sits still when he hears the line: "Yet she must die, else she'll betray more men." But he clenches his fist and fights back bitter tears. And he sees himself as one more man betrayed.

Neither he nor the yokel has heard the poet's music.

They hear only calls for action. Both are too close to the fire.

Bullough, however, liked his own art hot. He wrote: "What is, therefore, both in appreciation and production, most desirable is the *utmost decrease of Distance without its disappearance.*"[54]

In other words, you should come as close to the art object as possible in the sense of aesthetic distance without falling through the inner boundary of the aesthetic ring. But Bullough also understood—or, at least, asserted—that this "coming close but not too close" was a matter of great delicacy and required sensitive judgment and training and, perhaps, a certain artistic gift[55]:

> In theory, therefore, not only the unusual subjects of Art, but even the most personal affections whether ideas, percepts or emotions can be sufficiently distanced to be aesthetically appreciable. Especially artists are gifted in this direction to a remarkable extent. The average individual, on the contrary, very rapidly reaches his limit of decreasing Distance, his 'Distance-limit', i.e., that point at which distance is lost and appreciation either disappears or changes its character.
> In practice, therefore, of the average person, a limit does exist which marks the minimum at which his appreciation can maintain itself in the aesthetic field, and this average is considerably higher than the Distance-limit of the artist.

Picasso—Bullough implies—can stand close enough to his nude model to breathe her perfume. He can place his hands on her bare shoulders, and turn her to make the shadow fall between her breasts just so and not suffer the slightest loss of aesthetic distance or of artistic perspective. But our friend, the yokel, passing by outside, will break glass if he catches even a glimpse of the goings-on in the studio.

You can make the same point with bullfighting as, in fact, did Ernest Hemingway[56] in 1932 in his compelling book *Death in the Afternoon*. If you sincerely identify your-self with animals (Hemingway would have called you an animalarian), then the bullfight can be nothing to you ex-cept a brutal business which causes the infliction of great pain on innocent animals. According to Hemingway, the animalarians will see only horses gored and disemboweled and a bull speared and chased and finally put to clumsy death by a sword driven through his neck. Once in a while, the bull has a moment of luck and tosses a matador upside down and the animalarians turn humanitarians and wring their hands for the bullfighter struggling up on one knee in the sand and about to be smashed and skewered by a bull the size of a small car. But mostly the animalarians suffer for the animals, particularly for the horses.

These sufferers are far, far inside the inner boundary of the bullfight's aesthetic ring. In fact, they will stoutly argue that no aesthetic ring for the bullfight exists. The crude bullfight pushes against their noses; they can imagine no emotion coming from it except disgust, and emotions re-lated to disgust.

However, it is a clear fact that not everyone sees the bullfight as the animalarians see it. While not a sport in the Anglo-Saxon sense of the word, the bullfight for many years provided the people of Spain with grand entertainment spectacles vaguely analogous to what is provided us on fall Sunday afternoons by the National Football League. It is perfectly transparent that of those hordes of Spaniards who came routinely to the Corrida de Toros—the nobility, the shopkeepers, the sellers of oranges, the elegant ladies with dark combs in their hair—not all of them saw the bullfight as do the animalarians. Most of these spectators could maintain enough distance from the blood on the sand to put it—in Bullough's terms—"out of gear" with the horse or the bull or the man who did the bleeding. They saw the thing as something other than barbarism—entertainment, sure, but also they saw it as craft.

Hemingway asserted that one could go even further. One could, he claimed, distance oneself enough to see the bullfight as *art*. And this, Hemingway claims, and the associated descriptions and arguments to support it make up the core of *Death in the Afternoon*.

I will not comment on the correctness or incorrectness of Hemingway's "bullfighting as art" thesis. I have never seen a bullfight but I know what I am supposed to feel about them and what I have been taught to feel. Frankly, I believe it would take many exposures for me to separate these taught and accepted moral feelings from what I would truly feel, or could learn to feel, about the goring of the horses or the death of the bull. And I cannot see my potential education in this area to be worth the life of a single bull or the maiming of even one horse. My own bullfighting aesthetic position shall remain unknown to me, like the billionth digit in the expansion of π, or the exact structure of my cerebellum.

But Hemingway's book shows categorically that, even if bullfighting stands as far from art as Shanghai from St. Louis, one can write prose about it that itself is art. At least Mr. Hemingway could. And did in 1932.

Death in the Afternoon also illustrates directly Bullough's point that you can experience an object as art only if you are a certain distance from it; and you will have no chance of seeing it aesthetically if you come too close.

Mathematics, I am happy to say, is different from bullfighting. But it is similar in that a large subset of the population stands too close to it ever to have any chance at experiencing it aesthetically. I am speaking mainly of scientists and engineers and the others who, as I have noted, have had mathematics pressed against their noses in class after class

and course after course as something they must first master
before they can get to their real work, the science or the
engineering which they actually want to do. These people
—the scientists and the engineers—are unlikely to have
ever encountered the subject in any other manner. As Bullough put it, mathematics for them has "never been cleared
of the practical, concrete nature of its appeal." And because
this clearing has never—even for a single instance—taken
place, they cannot comprehend the existence of an aesthetic association with mathematics exactly as Hemingway's animalarians cannot imagine any connection between bullfighting and art.

This failure to impose sufficient distance between the
scientists and the engineers and the object called mathematics represents one of the great failures in mathematical education. Unfortunately, the present trend of undergraduate
mathematics instruction—particularly calculus instruction—which shows no slacking of the proliferation of meaningless techniques and trivial applications at the expense of
mathematical ideas and theory indicates clearly that the
educational process will continue to fail. And this particular failure is a failure of *underdistance*.

But aesthetic failure can also come from too much distance. Out there beyond the aesthetic ring—where we find
our observer in Figure 12—there exists failure of a different sort. Beyond the aesthetic ring lives ice rather than fire.
And when Bullough describes aesthetic failure in this cold
world, he uses adjectives like "improbable," "artificial,"
"empty," and "absurd."

You commonly find too much aesthetic distance associated with "contemporary" or "abstract" art. Squiggles on
canvas, stacks of driftwood, random atonal violin notes are
routinely rejected as art by the majority of the population
no matter how often they are told the objects are not what

they seem but are, in reality, serious works of painting, sculpture, or music. The objects are, in an aesthetic sense, simply too far from the typical observer who sees them as cold and remote and who reacts to them mostly with indifference. These highly distanced works of art are often deliberately made as antirealistic as possible in order that those who make them can assert that the thing is, in fact, something exalted and capable of being appreciated only by those who come to it willing to contemplate, to study, and to learn.

This deliberate distancing of art by artists may be neither improper nor incorrect. Ortega y Gasset—as I read him—praised, at least partially, certain aspects of this tendency which he labeled "dehumanization." The idea is to eliminate the human characteristics of art—like realism in painting and story line in literature—on which the average person invariably depends to make the art palatable. The result, said Mr. y Gasset, will be an "art for artists," which is not necessarily bad. He wrote[57]:

> Even though pure art may be impossible there doubtless can prevail a tendency toward a purification of art. Such a tendency would effect a progressive elimination of the human, all too human, elements predominant in romantic and naturalistic production. And in this process a point can be reached in which the human content has grown so thin that it is negligible. We then have an art which can be comprehended only by people possessed of the peculiar gift of artistic sensibility—an art for artists and not for the masses, for "quality" and not for hoi polloi. That is why modern art divides the public into two classes, those who understand it and those who do not understand it—that is to say, those who are artists and those who are not. The new art is an artistic art.

Mr. y Gasset, so far as I know, was not expert in mathematics. Nor did he concern himself with matters mathematical. But the above paragraph describes almost exactly the attitude—often unconscious and implicit—of modern-day mathematicians toward their subject and its relationship to people outside the mathematical aristocracy. All you need do is substitute "mathematics" for "art" and y Gasset is saying

> We then have a mathematics which can be comprehended only by people possessed of the peculiar gift of artistic sensibility—a mathematics for mathematicians and not for the masses, for quality and not for hoi polloi.

Then you have a point of view agreeing simultaneously with Poincaré and his notion of "innate" mathematical aesthetic sensitivity and with the contemporary research mathematicians who have clearly created a mathematical world for themselves alone. And while this type of attitude and practice may not be bad for art—as Mr. y Gasset assures us—it is devastating to mathematics.

Bullough's ideas, I believe, contain much that is applicable to mathematics and to mathematics teaching. Even his most basic notions—those of aesthetic distance and the associated concepts of "overdistancing" and "underdistancing" help us to understand the different perspectives of mathematics held by the scientists on the one hand and the humanists on the other. My extension of Bullough's ideas yields a model in which these groups are separated by the mathematicians. Thus, mathematics may be seen as either the *barrier* between or the *bridge* joining the famous two cultures of C. P. Snow.

We will return to this point later. For now, we'll content ourselves with Bullough's notions and their applicability to the concept of mathematics as art. As we proceed, it will be helpful to keep in mind Figure 13 which shows mathematics as an art object labeled M and its aesthetic ring where the mathematicians live—drawn shaded and dark, as it appears to most of us who glimpse it only occa-

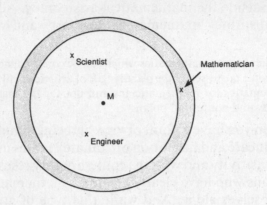

FIGURE 13. The aesthetic ring of mathematics.

sionally and furtively through the narrow, dim slits in the cloister walls.

INTUITION QUIZ ANSWERS

All answers are *false*. Questions 1 and 2 are popularizations of Bertrand Russell's famous paradox which played a key role in his attempt to reduce all of mathematics to a few fundamental logical principles. You can find the paradox and related matters discussed in the little book *Gödel's Proof,* by Ernest Nagel and James R. Newman.[58] In the case of problem #1, if the barber shaves himself then he belongs to the set of people who do not shave themselves and you have a contradiction. On the other hand, if problem #2 were true then the barber belongs to the set of people who shave themselves and, consequently, are not shaved by the barber. Again there is a contradiction. The way out of the paradox is to deny the existence of such a village and such a barber. The way out of Russell's paradox in the general case is similar but the consequences for mathematics and, hence, for the world are more profound.

Problem #3 is an illustration of the famous "Birthday Problem" which is covered in most elementary courses in probability. An intelligible discussion can be found in one of Martin Gardner's puzzle books.[59] Actually, the probability is better than 1/2 that some 2 out of 23 randomly selected people will have been born on the same day of the year. For 50 people, the probability that at least two will have the same birthday is about .97 and the odds for a fair bet are 33 to 1.

An explanation of the falsity of #4 requires knowledge of the theory of maxima and minima of functions of two variables. The "explanation" amounts to producing a surface and a point on it such that the point looks like a peak if you move away in any straight line but not if you move on some curve. An example of such a surface appears in the book *Advanced Calculus*, by Angus E. Taylor.[60]

If the couple in #5 has two children then the various possibilities are *gg*, *gb*, *bg*, *bb*, where the symbol *bg* denotes "oldest child a boy and youngest child a girl," and the other symbols are similarly defined. If we assume that boy births and girl births are equally likely then so are each of the four boy–girl combinations. Since there are four possible combinations, the probability of each is 1/4. However, if we know that one of the children is a girl then the possibility *bb* is eliminated. Thus, the possibilities become *gg*, *gb*, *bg*, each of which is equally likely. The probability of *gg* now becomes 1/3.

Problems #6 and #7 are also discussed by Martin Gardner.[61] It turns out that, in this situation, Carter's best

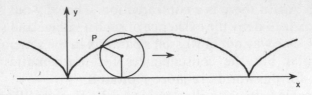

FIGURE 14. The cycloid.

strategy is to fire in the air and let Adams and Brown fight it out—a nonintuitive procedure which may say something profound about the conduct of international affairs.

The last problem is classical and famous. It is called the "Brachistochrone Problem" from the Greek words for "shortest time." The problem was proposed, and solved, by John Bernoulli in 1696. It was also solved by John's brother, James Bernoulli, and by Newton and Leibniz. The descent curve of *minimum distance* is, of course, the straight line segment joining the two points. But the curve which produces the *minimum time* of descent is an inverted *cycloid*. This curious curve can be geometrically described as the path traced out by a fixed point on the circumference of a circle as it rolls along a straight line (see Figure 14). A technical (but elementary) discussion of these matters can be found in the seminal calculus book of George B. Thomas.[62]

CHAPTER 6

Aristocracy

ne of the facts of college life which often escapes the attention of outsiders is the pervasiveness of academic conformity. Everyone knows that students look alike. Choose any college and compare its 1955 yearbook with the yearbook for 1969. You will find that the pictures from 1955 show students in short haircuts wearing jackets and ties or sweaters and pearls. In 1969, both the men and the women have hair down to here and wear torn jeans and faded shirts ornamented with feathers and beads. The 1955 students will be so dissimilar from the 1969 students as to have come from different worlds. But the students in 1955 all will look alike as did the students in 1969. On any campus, at any given time the students are, from ten feet away, as identical as snowflakes.

The fact that students think alike is less well-known but does from time to time manage to reveal itself off campus. This theme recurs in Allan Bloom's *The Closing of the American Mind.* Bloom, in fact, says in the first sentence following the preface: "There is one thing a professor can be absolutely certain of: almost every student entering the university believes, or says he believes, that truth is relative."[1] Bloom finds this scary and so do I. But more frightening is the conformity of thought of students leaving the university: a conformity primarily coming from the local homogeneity of their professors.

One of the games which campus visitors enjoy consists of trying to identify, in the faculty dining room, the various academic disciplines based on the mode of dress of the persons seated at particular tables. The game depends on the high statistical validity of two facts: (1) faculty from any particular department will dress alike, and (2) most persons seated at a particular table will be members of the same academic department. The game varies from campus to campus because the local standards of dress vary although ordinarily not by much. At my university, the best dressed faculty—in the traditional suit and tie sense—are the faculty from the departments in the college of business. Next come the engineering faculty with the industrial engineers heading the list. In the college of arts and sciences, the dress varies significantly from department to department with the historians near the top. On the other side—in terms of blue jeans and open shirts—there exists almost a three-way tie among the sociologists, the psychologists, and the mathematicians. When you play the faculty dining room game at my place you have no trouble distinguishing the mathematicians from the accountants. But you will have trouble telling them from the psychologists.

Please understand that I am not making value judgments. I am simply stating the fact that, in the university, members of a particular academic department tend to dress alike. Of this, there is no doubt.

The mode of dress, of course, may vary with time. Not long ago, a colleague from the mathematics department came to my dean's office to complain about the administration. At one point he mused that perhaps he should become a dean on the grounds, assuming, I suppose, that things would then be done right. But then he looked down at his open shirt and said: "But to become a dean I'd have to start wearing a tie. That's what you did."

I said nothing to that. I rose and pulled two yearbooks off my shelf, the most recent book and the one from a quarter century past. I opened each book to the page which held the mathematics department group picture. The old photograph showed each male faculty member, including the two of us, wearing a suit and a tie. I was not in the most recent photo but he was. In that picture, nobody wore a tie and the mode of dress was casual enough for a picnic.

I placed the two photographs in front of him. He looked at them. "What's your point?" he said. I said nothing to that either.

The conformity of dress and behavior shown by university professors consists primarily of departmental conformity. Hazzard Adams in his book *Academic Tribes* writes: "The fundamental allegiance of the faculty member will be to the smallest academic unit to which he belongs."[2]

This "smallest unit" is normally an academic department and the "loyalty" normally extends to a faithfulness to departmental standards of dress and behavior. This faithfulness manifests itself particularly in mathematics departments.

But the mathematicians at research universities go far beyond simple outward manifestations of homogeneity. Academic research mathematicians have developed a set of values, behaviors, and attitudes that sets them apart from their colleagues in other departments as sharply as the English nobility stand apart from street people in Soho Square.

We need to examine this academic class distinction to ascertain whether or not its existence acts as an obstacle to improvement in mathematics education and to increasing the awareness among educated people of the scope and the importance of mathematics research. For two facts are clear:

- mathematics education is a failure
- mathematics research is unknown

If, after reading the daily newspaper accounts of the dismal mathematical performances of American students when compared to students of other nations you are not convinced that mathematics education has failed, you need only listen (as we have noted) to former presidents Lynn Steen and Leonard Gillman of the Mathematical Association of America describe calculus instruction as "failed" and "deplorable." As for research, it may be that mathematicians publish 25,000 papers per year, but no one outside of the community of research mathematicians has seen them, read them, or even heard of them.

Something else is clear. The academic *subject* which the mathematicians have in their possession, as the Tower Yeomen have the crown jewels, cannot be held accountable for the defects of either research or education. This thing called mathematics is too lovely and too valuable to be at fault. The blame lies not in mathematics but somewhere else—probably in the way the subject is kept and shown by those who have the keys to the jewel room. Perhaps the way to bring mathematics out into the sunshine is to first bring out the mathematicians. But before this possibility can

even be considered we need to know exactly where the
mathematicians are now. We must learn the characteristics
of the society of mathematicians. What are they like?

Morris Kline had no hesitation in putting forward his
notion of their characteristics. In his book *Why the Professor Can't Teach* he wrote[3]:

> Mathematicians have always constituted a clannish, elitist,
> snobbish, highly individualistic community in which status is
> determined, above all, by the presumed importance or original
> contributions to mathematics; and in which the greatest re-
> wards are bestowed upon those who, at least in the opinion of
> their peers, will leave a permanent mark on its evolution.

I do not know if research mathematicians have always
been this way but my own experience is that for the past
thirty-five years, Kline's description has been highly accurate. Professor Kline certainly did not intend to be kind to
mathematicians when he wrote the above sentence. But,
with the possible exception of the word "snobbish," the
description contains no pejorative connotations. It might
be snide to describe a collection of people as "clannish"
and "elitist" but, on the face of it, it is not deriding. A
problem occurs only if Kline's description is accurate and if
these characteristics are simultaneously such that, having
them, the mathematicians cannot properly do their job. Or,
in fact, if by having these characteristics, the mathematicians fail to understand *what their job ought to be.*

Kline's book, incidentally, was a scathing attack on the
values and practices of the community of research mathematicians. The book received considerable attention and
was favorably reviewed in the *New York Times* and the
Wall Street Journal. Because of this attention and because
Kline himself is a distinguished mathematician you would
expect a clamor of response activity from the community of
research mathematicians—response either as rebuttal or as

reform. But almost none came. The mathematicians re-garded Kline's book—if they regarded it at all—as inferior criticism and simply went on with business as usual. And this lack of response gives Kline's use of the word "snob-bish," at least, some small credence. More important, the fact that no one else in the academic world—administra-tors or other faculty—even *expected* the mathematicians to respond to Kline illustrates a point to which I will return shortly: mathematicians occupy a position of privilege within the academy.

The mathematician P. J. Hilton[4] did write a response to Kline. But Hilton's article came in response to an invita-tion of the editors of *The Mathematical Intelligencer*, a journal which had earlier published excerpts from Kline's book. He spoke presumably, only for himself. (Both posi-tions are, in my view, extreme. Kline essentially says that good research inhibits good teaching while Hilton comes close to asserting that good research implies good teach-ing.) And there were one or two other individual responses —an article or two and a letter here and there. But, by and large, the research mathematicians went on as before. They were untouched by Kline's book as they are untouched by everything which does not somehow involve the creation of new mathematics.

Timothy O'Meara is Kenna Professor of Mathematics and Provost at the University of Notre Dame. In October, 1985, he gave the keynote address at the National Chair-man's Research Colloquium for the Mathematical Sciences in Washington D.C. His talk, "Strategies for Enhancing Resources in Mathematics," was published in the Notices of the American Mathematical Society. In the talk, O'Meara described asking several nonmathematicians the question: What do *you* think of mathematicians? Here's what his associate provost, an ethicist, said[5]:

They are self-contained; they presuppose that what they are
doing is relevant whether or not anyone else thinks it is; they
have a great tolerance for individuals; they consider neither
social conformities nor appearance to be of much importance;
they reach a high level of competence at an early age; after that
a certain boredom sets in which seems to affect the way in
which they teach.

This description seems entirely consistent to that provided by Morris Kline; yet it introduces something new,
namely, the *boredom* mathematicians associate with
teaching.

There is no doubt that large numbers of research mathematicians find teaching—particularly calculus teaching—
boring. There's nothing intrinsically wrong with this. One
cannot help being bored. Lawn mowing bores me. But I
mow anyway. And I mow well. What is wrong is to allow
the boredom to affect the work. O'Meara's associate provost is saying that—in the job of teaching mathematics—
this often happens.

Mathematicians may allow the boredom to show
through because teaching is not significant to them.
Whether they do it well or do it poorly is often a matter of
indifference. Teaching, for many research mathematicians,
simply is not part of the university reward system. Speaking of mathematics teaching, Morris Kline says flatly:
"Teaching just does not count in universities. Of course
administrators deny this."[6]

The statement is extreme. But if the word "just" is
replaced by "often," it becomes accurately applicable to
research universities.

I am hopeful the mathematicians will one day return to
teaching. When they do, they will have to struggle very hard
against allowing the unavoidable boredom of giving yet another calculus lecture affect the way they in fact give the

lecture. The topic may be derivatives of trigonometric function and it may be, to the professor, too old and tedious. But to the students it is brand new. The teacher's job is to present it freshly.

Here's what John Barton said in *Playing Shakespeare*[7]:

> The words must be *found* or *coined* or *freshminted* at the time you utter them. They are not to be thought of as something which pre-exists in a printed text. In the theatre they must seem to find their life for the first time at the moment the actor speaks them.

Exactly. Replace "theatre" by "classroom" and "actor" by "teacher" and you have a two-sentence handbook on effective mathematics teaching.

Near the end of his talk, Timothy O'Meara returned to the question of what mathematicians are like. He wanted to identify the academicians that were closest to mathematicians. He said[8]:

> My instincts kept telling me that they were theologians. But a piece was missing from the puzzle. Last night my wife and I talked about it, and we found the answer. The difference between mathematicians and theologians is that theologians have more articulate spokesmen, they have more gurus. It is the *cloistered* theologians who resemble us most.

O'Meara's analogy is not bad. It works even better if you think of cloistered monks rather than "cloistered theologians." Mathematicians certainly are cloistered in the sense that they work in isolation from their nonmathematical colleagues. The work they produce is as intellectually inaccessible to those colleagues as fourteenth-century Latin religious documents were inaccessible to the local peasant farmers. (Each academic department stands—in some sense—separate from the rest of the university. But, for the mathematicians, the separation is extreme and almost total. No one in the university—with the possible exception of a tiny band of theoretical physicists—has any real notion of what mathematicians do.) Mathematicians are cloistered also in the sense that they are unaffected by what goes on outside their walls. For monks, the walls were made of stone and they kept the world away. The mathematicians are secluded by walls formed of mystery—the mystery associated with work that is unknown and unintelligible to administrators and other professors and yet continues to be deemed worthy of support. And, make no mistake about it, the walls of mystery serve the mathematicians exactly as the walls of stone serve the monks: they keep the wind out.

Winds of change and reform may blow through the secular world but the monks go on writing theology in Latin script. The same winds blow through the academy and the historians and the chemists and the engineers are buffeted and forced to contend. But the mathematicians remain untouched and continue writing mathematics on yellow pads and grey chalkboards. Morris Kline may demand "mandatory reform" in mathematics teaching and he may reach the readers of the *New York Times* and the *Washington Post* but the mathematicians will not hear him. And Gillman and Steen and the others may contend that the basic course taught each year to one million students is "scandalous" and a "failure" but the mathematicians will not be affected. The mathematicians are in out of the cold—untouched; blow, winds and crack your cheeks!

O'Meara's "theologian analogy" fails partly because academic theologians do not write theology. Rather, they provide a kind of scholarly criticism of the historical and psychological theology written by someone else. Mathematicians, as we've seen, are not critics. In fact, since the publication of Hardy's *Apology,* mathematicians have considered criticism to be work fit only for "second rate minds." Mathematicians *write* mathematics as monks write theology.

However, the monk analogy fails also because the monks recognize the existence of a higher authority and it is for the glory and knowledge of this supreme being that they lock themselves away and contemplate and write. Mathematicians write mathematics because they like it and they find it beautiful. Mathematicians recognize no authority except one another. A better analogy than theologians or monks exists and we will come to it shortly.

But, before we leave Timothy O'Meara, let's notice the response he got from the mathematicians to his keynote address. O'Meara's task, he said, was[9]

> to use my perspective as mathematician and provost to reflect on some of the perceptions of mathematicians which are held by the university community at large in order to identify some of those forces *internal* to the mathematics community which are acting as obstacles to future development.

In other words, O'Meara was to talk about how mathematicians are *perceived* and how that perception can be harmful. Following the talk were questions and answers. Five of these questions were published in the *Notices of the American Mathematical Society.* Three of them were concerned directly with research and support for research. One question had to do with the number of semesters which should be required for college mathematics. None of these, of course, had anything to do with the subject of O'Meara's talk. It seems that, no matter what you say to mathematicians, what they say back to you will be about mathematical research.

O'Meara's purpose was partly to advise mathematicians as to how they might improve their position among other departments and disciplines in the competition for scarce funding resources. With regard to his advice, Provost O'Meara was asked a fifth question: "Instead of telling us what to do, why doesn't someone speak to those dumb deans and provosts?"

PRIVILEGE

The walls of mystery that surround mathematicians provide them with privilege rather than prison. The walls defend them. Almost no one touches them, neither parents nor provosts nor hordes of dumb deans.

Alfred Adler wrote in the *New Yorker*[10]:

> Perhaps mathematicians, lacking the imagination to appreciate the scope and sophistication of the outside world, confuse minor success with real achievement and are satisfied with it. Then, too, they seldom recognize failure when they are confronted with it; rather, they tend to think of it as simply one more betrayal by a society that usually patronizes them while elevating armies of patently inferior claimants. In the academic world, on the other hand, mathematicians often enjoy rewards that they do not merit. They are engulfed by admirers from the departments of philosophy and the social sciences . . .

Truly, mathematicians "often enjoy rewards that they do not merit" and these rewards result primarily from the cloistering of the mathematicians within the walls of mystery. Nonmathematicians in the academy simply do not know what mathematicians do. However, they believe the work to be valuable exactly as the fourteenth-century peasants believed the monk's work valuable. To the nonmathematicians, all mathematics is deep and unintelligible. Its value is accepted as an article of faith.

The mathematicians' "admirers" come not only from philosophy but from all disciplines which strive toward quantification. The physicist L. T. More says[11]:

The supreme value of mathematics to science is due to the fact that scientific laws and theories have their best, if not their only complete, expression in mathematical formulae: and the degree of accuracy with which we can express scientific theory in mathematical terms is a measure of the state of a science.

Thus, according to More, you can "measure" the development of a science by noting how mathematical it has become—the development being directly proportional to the mathematical sophistication. And More uses "science" in a general sense. He writes[12]:

Thus it is possible to classify sciences according to their development from the accumulation of statistics of phenomena to the generalization of these phenomena in comprehensive and rigorous laws. In such a classification, sociology or the study of existing society occupies the lowest rank, since true laws can be derived only from actions whose consequences are known. Sociology therefore attempts to found its laws on the study of history, the study of past society; history must in the same way rely on psychology, which deals with the actions of individuals of society; psychology relies on biology; biology on chemistry; chemistry on physics; and physics on pure mathematics.

Pure mathematics, therefore, lives at the top of the academic pecking order and even the shallowest and most trivial of mathematicians inherits an unearned quota of admiration. And the admiration is accepted. Alfred Adler says: "Mathematicians are too vain to assess such admiration at its true worth."[13]

In the academy, many consequences follow from the existence and acceptance of this admiration and from the perpetuation of the mystery of mathematics. One consequence, of enormous significance, is the great degree of freedom from external control and evaluation enjoyed by mathematicians. Nonmathematicians, both professors and administrators, feel illiterate and intimidated in the presence of mathematics and believe themselves incompetent to bring to bear on mathematicians standards they routinely apply to other colleagues.

You see this absence of control and evaluation clearly in the case of the "deplorable" calculus teaching situation. Here, a teaching system which has been marked "failure" and "scandalous" by almost every observer continues because the mathematicians *alone* want it to continue and no one in a position of academic or administrative leadership has the mettle and spirit to challenge it. Only in mathematics, where esteem follows from mystery, could such a situation occur. Were the historians, for example, to try and present a basic required course in as offhand and slipshod a manner as calculus is often presented, they would immediately face declining enrollments, faculty censorship, and possible loss of budgetary departmental positions. But calculus is required—for science, for engineering, for anything technical. Substantial enrollments are guaranteed.

But—beset with admiration for a subject they do not understand—administrators "sign off" for mathematics. The mathematicians go their way. The mystery deepens. The cloister walls thicken.

To be fair, I must point out that the mathematicians did not originally ask for special treatment. They have now, after these many years, come to expect autonomy with respect to evaluation of teaching and research. But in the beginning they did not request privilege. It just came their way.

What happened was that American mathematicians were transported, in a discontinuous quantum leap, from obscurity to national prominence following the launching of *Sputnik I* in 1957. Overnight, mathematics and mathematics research became essential to the national interest. The Russians were beating us in space. And—the nation thought—they were beating us with mathematics. The headiness of national attention and the excitement and rewards of research turned the mathematicians away from teaching. Large calculus lecture sections appeared on almost every major university campus—justified, ostensibly, by increased enrollments but, actually, by the need to keep teaching loads low in order to free up professors for even more research. As time passed, the notion of teaching changed from that of a meaningful part of a mathematician's job to almost an irrelevancy. The idea of mathematics as part of a liberal education vanished and elementary calculus became reduced to a grab bag of techniques only partly understood and immediately forgotten. Students came away from college courses with their dislike for mathematics confirmed. And they came away with no notion whatever of the nature—and often even the existence—of this thing called *research* which seemed to occupy so much of their professors' time and attention.

After a while no one cared. Fear of Sputnik faded. The outside world shrugged and went about its business.

Here's Adler again[14]:

> And, finally, there is the non-mathematical world, in which the mathematician appears unable to find success, and which at almost all points accords the mathematician a monolithic indifference. So there is no way out for mathematicians; there is no place for them to turn but to other mathematicians and inward on themselves.

Yes. And the inward turning is now a quarter-century established. For all this time, the mathematicians have talked mathematics only to each other and about mathematics to no one at all. What we hear in the academy about mathematics comes mostly from the scientists and engineers who use mathematics as a tool and who, being far too close to the subject in terms of aesthetic, see in it no aesthetic value. But, surprisingly—in spite of the fact that the scientists say mostly wrong things about mathematics and the mathematicians say nothing about it at all—some memory of the intellectual value and the importance of mathematics lingers in the academy. Some mental relic of things learned long ago or else some instinctive feeling for the appropriateness of mathematics remains in the minds of administrators and nonmathematics faculty. Thus, in spite of its mystery (and maybe because of it) academicians continue to believe mathematics and mathematics research to be worthy of support.

The mathematicians turned inward on themselves and walled out the world. Being cloistered, they could do mathematics without the bother of dealing with the academic world and its bureaucracy in the way others had to deal with this world. But they still needed the *support* of that bureaucracy. Academic mathematicians need supplies and salaries, libraries and offices. They need tenure-track faculty positions and they need paid people to fill them. Fortunately—for the mathematicians—the support continues. But for how long?

You can stretch the "cloistered monks" metaphor only so far. One of the characteristics of the ancient monasteries was the attempt to be self-sufficient. Some of them grew their own food and made their own clothing. The monks worked only part of the day at their theology; the rest of the time they worked at merely staying alive.

There is not the remotest analogy here with research mathematicians. Mathematicians have turned inward to be sure. And they have cloistered themselves away from the rest of the university. But there is no pretense at self-sufficiency. Mathematicians do mathematics at the sufferance of others.

When you cloister yourself away, as the mathematicians have, and yet remain completely dependent on the outside, as the mathematicians are, you do not become monks. When you are cloistered and yet simultaneously supported by outsiders you become privileged. You become part of a distinct class. And such a class is called an *aristocracy*.

NOBLESSE OBLIGE

The mathematicians did not ask for privilege and mainly do not even know they have it. Since they turned inward in the early sixties they became first uninterested and then unable to understand the subtleness of academic bureaucracy. They have now been too long inside the walls, too far removed from the struggle for scarce academic resources, to compete with other disciplines for their share of these resources. They endure because they are not *expected* —as are the others—to compete. Adler says this of the mathematicians:

> However, their aversion to any but extreme and speculative positions has caused them to forfeit even the modest power and influence due them in bureaucratic affairs, where, consequently, mathematics at all levels is much less influential than any of the other sciences and much less influential than its scientific importance and its procedural virtues warrant.

Adler believes, I think, that there exists something in the very nature of mathematics which causes the men and women who study it for its own sake "to forfeit" relationships with the world of affairs. For he goes on to say[16]:

> All this contrasts vividly with the achievements of mathematicians when they do mathematics: meaningful results of brevity and simplicity, accomplished by an insistence on total rationality both in hypotheses and proofs. The professional restraints

are so severe that the reaction is too powerful. As soon as the bonds are loosened, mathematicians adopt careless procedures that, together with a vast self-esteem and a conviction of intellectual superiority, cause them to overlook crucial aspects of whatever they are doing—to lose the mental self-control necessary to almost every successful human endeavor.

Adler's conclusions that mathematicians have a "vast self-esteem and a conviction of intellectual superiority" are mainly correct. It may be that mathematicians lose "mental self-control" when the rationality "bonds are loosened" the way a fundamentalist preacher's kid loses control once he is out of the house. But, if it is true, all it says is that mathematicians lack discipline when they deal with something other than mathematics. It does not tell us that external "careless procedures" are the inevitable consequence of a commitment to mathematics.

What the mathematicians have in common is an understanding of, and an appreciation for, the aesthetic quality of mathematics. This common and unique appreciation for the elegance of mathematics defines the mathematician's aristocracy. It is an *aristocracy of elegance.*

Along with Euclid, the mathematicians have "looked on beauty bare."[17] Its brightness may have made them blind to the world outside and the blindness makes them stumble and lose control. Perhaps mathematicians react to the beauty of mathematics the way the Trojans reacted to the beauty of Helen. Helen simply *was.* She launched no ships. But men reacting irrationally to her face launched a thousand. And they bloodied other men for a decade[18]:

> Helen must needs be fair,
> When with your blood you daily paint her thus.

Adler, you will remember, said that mathematicians are "too vain" to properly assess the "admiration" of outsiders. He is mainly correct in this observation. Mathematicians simply are too cloistered to realize that, in the academy, they constitute a privileged class. Generally, the mathematicians know far too little of the organization and governance and bureaucracy of the university to which they belong to realize that, in matters of promotion and evaluation and teaching expectations, they receive special treatment. But they do. The other faculty—those who participate regularly in university affairs—*know* they do. What follows is the inevitable development of attitude that inferior classes always have for the superior class: the academic nonmathematicians view the mathematicians with envy and resentment. They envy the privileges allowed the mathematicians in terms of autonomy with respect to the setting of teaching loads and the evaluation of teaching and research. And they simultaneously resent the mathematicians for being privileged while they are not.

The mathematicians, on the other hand, would have everyone (and hence, no one) privileged. Suppose you took a mathematician, pried him away from his desk, and led him by the hand through the normal university budget process and promotion process and then said to him: "Look. You see all these procedures, all these checks and balances, all this paperwork and accountability. This, Mr. Mathematician, is what everybody else in the university must go through to justify teaching slots, and promotions, and graduate student support. Everyone, that is, except you guys. When you want to promote someone you just *say* he is a good teacher or a good researcher. Nobody expects the mathematicians to produce paperwork. No one expects you to have to justify anything. You live, Mr. Mathematician, in a world apart. A privileged world."

"Exactly, Mr. Bureaucrat," the mathematician replies. "But it should be this way for *everybody*. We are professors.

And it is our job to do research. Our business is new knowledge, not meetings and memoranda. All of us—every professor in every discipline—need exemption from these nonsensical university procedures so that we can be free to do our work."

When the mathematicians turned inward a quarter of a century ago and walled out the world, they took inside certain philosophical points of view; one of which is that freedom from restraint releases goodness. Regarding this notion, it is well to remember Bertrand Russell: "To believe that liberty will ensure moral perfection is a relic of Rousseauism, and would not survive a study of animals and babies."[19] Nor would it survive a study of mathematicians.

Mathematicians do not ask me for advice. So far as I know, mathematicians ask no one for advice except other mathematicians and, then, only on matters directly pertaining to mathematics. However, were they to ask me, I would tell them three things:

1. Choose some leaders who have the skills and the experience to deal with a nonmathematical world. To do this you must realize that the choice of leader may have to be made independently of the person's ability to do mathematics research. (In fact, if Alfred Adler is correct, it is highly unlikely that there exists a first-class mathematics researcher with the requisite skills to obtain a rapprochement between the mathematics aristocracy and the outside world. I believe Adler to be incorrect on this point; the mathematicians' disinterest in academic leadership may be just an extension of Hardy's scorn for critics, publicists, and anything noncreative.)
2. Forget forever the silliness that good research *im-*

plies good teaching. Everyone who has looked carefully at the collection of good mathematics teachers
and the set of good researchers knows this is not
true. Mathematics research can, and should, be justified on its own merits. The irrational attaching of
teaching to research degrades both. Certainly, there
is a positive relationship between research and
teaching. But the correlation is statistical and has
no *prima facie* value in individual cases. It does not
follow that Professor Deep is a good teacher *because* he is a good researcher. Each time this argument is advanced, the credibility of mathematicians is reduced.

3. Accept the notion that you occupy a position of
privilege within the academy and associate with it a
concomitant obligation, a notion of noblesse ob-
lige. The basic obligation should be to widen mathematics' aesthetic ring to bring in a sizeable portion
of both the scientists and the humanists. This
means that you must return to the old notion of
mathematics as part of a liberal education and you
must go back to the time prior to the walls when
university teaching was a serious part of a research
mathematician's job. It also means that you must
abandon the G. H. Hardy notion which associates
mathematical exposition with failure. On the contrary, it is vital that you identify a collection of people who can explain mathematics and mathematics
research to nonmathematicians and that you reward them for doing so.

Before the mathematicians could accept the advice of
number 3, they might have to select the leaders I describe in

number 1. They have been cloistered for a long time. Teaching and mathematics exposition have long ago been chucked over the wall. Before the mathematicians can rationally discuss these notions they may need someone to tell them how to discuss them. It may be necessary for them to listen seriously to a nonmathematician.

It would be well for the mathematicians to find such leadership. For the aristocracy of elegance (as do all aristocracies sooner or later) has become dusty and tired. Twenty-five years of inbreeding have produced a cloister of mathematicians who are of almost identical professional manner and mindset. Other than their unique appreciation for the elegance of mathematics, what you notice most about them is their almost total homogeneity. They pretty much look alike and think alike. And, as we have seen, they have allowed all of the humanistic aspects of mathematics to be removed from their elementary courses in favor of the inclusion of techniques.

Students do not learn that mathematics has developed exactly as have all other arts and sciences, as the result of great efforts made over long periods of time by real men and women. They hear of Weierstrass, for example, but they learn nothing of his student life at Bonn where he worked at beer drinking and fencing with the same dedication he later brought to mathematics. Nor do they learn of his pupil Sonja Kowalewski, and their deliciously scandalous relationship when he was fifty, she twenty-five, dazzling, and on her way to mathematical greatness. Galois Theory is taught in mathematics classrooms, but nothing is taught of the sad and romantic Evariste Galois, writing mathematics by candlelight, scribbling in his margins, "I have not time; I have not time," trying desperately to hold back the dawn. But the dawn came and Galois was—as he foresaw—defeated on a "field of honor," shot dead in a duel at age 21. These are real people. But they do not appear in mathematics classrooms.

Students see the human side of mathematics only through their professors. Evidently, the students do not want to be like their mathematics professors.

The mathematicians, like all aging aristocracies, desperately need the fresh blood of new ideas. But the ideas they most need are ideas associated with teaching, with the explication of mathematics, and with the relationship of mathematicians to the outside world. But mathematicians tend not to be receptive to these ideas. In fact, inside the cloister where the mathematicians sit facing inward in concentric circles, the talk is only of doing mathematics and the needed ideas have become unthinkable thoughts. Like sunshine, they will come—if they come at all—from the outside.

But come they must. Or else the party's over. For the mathematicians cannot continue as they are. Aristocracy or not, no class that is simultaneously privileged and unobligated can long endure. And the aristocracy of elegance has already lived more than twenty-five years. Walling themselves in, the mathematicians have smothered mathematics away from six generations of college students. It can't continue indefinitely.

> I know you all, and will awhile uphold[20]
> The unyoked humor of your idleness.
> Yet herein will I imitate the sun,
> Who doth permit the base contagious clouds
> To smother up his beauty from the world,
> That, when he please again to be himself,
> Being wanted, he may be more wondered at
> By breaking through the foul and ugly mists
> Of vapours that did seem to strangle him.

PARADOX

From the outside, college professors are perceived politically to be a collection of cookie-cutter liberals. They are thought to live on the far left of the political spectrum and to possess uniformly homogeneous views regarding the role of government as an instrument of social reform. And, as the American electorate—at least in recent national elections—has tended to become more conservative, this outside perception has contributed to the erosion of public confidence in higher education.

The perception is mainly correct. To be sure, there are college professors whose political views lie far to the right of center. The term "conservative professor" is not—as many academic critics would have you believe—an oxymoron. Still, the prevailing professorial political point of view—particularly in the major universities—is largely liberal.

Ordinarily, this is of no consequence. The political persuasion of any collection of college professors has academic significance only if the politics works its way into the classroom. So long as a professor keeps his political views separate from his role as a teacher of impressionable youngsters, what he believes or does not believe is of no more concern to me than are the political leanings of my dentist. Unfortunately, a professor's politics often does appear in the classroom resulting in either a political indoctrination of the students or in a further diminishing of the students' respect for the educational process. Either way, the intrusion of professorial politics into the classroom is unfortunate at best and, at worst, unacceptable.

Mathematicians, here, are clean. In all my student years of sitting in mathematics classrooms, I cannot recall a single instance of a professor using class time to assert political views. And, in subsequent years, I have never come upon—during one of my surreptitious visits to calculus lectures—a mathematician who talked in class about anything other than mathematics.

This does not mean that mathematicians fail to hold uniform political views. What it means is that, when they face students in the classroom, they talk about what they are supposed to talk about: mathematics. In fact, mathematicians are generally—I believe—even more to the liberal left of the political spectrum than are most of their academic colleagues. Alfred Adler says[21]:

> Nor do mathematicians distinguish themselves by their political activities at any level. They have been known for their radical positions, usually on the left but sometimes on the right—positions defended emotionally and often irrationally.

I believe Adler's characterization of mathematicians as being "usually on the left" to be a massive understatement. My own experience has been that, if you select an arbitrary research mathematician from any of our major universities, he will almost certainly be a political liberal. The conservatives form—as the mathematicians say—a set of measure zero.

I have no doubt of the dominance of liberals in the collection of research mathematicians. I have seen it for thirty years. But, I am also certain it is of no academic consequence since the mathematicians do not bring their politics into their classrooms. The mathematician Professor Deep may live to the left of Leon Trotsky but, even when in his darkest despair, humbled in front of hundreds of students and struggling with the limit of $(\sin x)/x$, he does not bail himself out by reading from the thoughts of Chairman Mao.

These political views are of no academic consequence but it is interesting to contrast them with the characteristics and the behavior of mathematicians when they are considered as part of the semipolitical setting in which they are located professionally: the American university. When you make this examination you find that the externally liberal mathematicians paradoxically form, within the university, its most academically conservative professorial subset.

For example, the mathematicians as political liberals feel comfortable with the general notion of government as a social instrument. They welcome—as, indeed, do many other Americans—the notion that you can, through governmental regulation, enforce significant changes in behavior with respect to racial discrimination and you can change social attitudes toward abortion and sexual preferences. Moreover, they understand—if not welcome—the necessity for government to increase in size and complexity in order to bring about these changes. But these standard liberal notions are in no way carried over into their own activities as part of the university community—nor toward initiating and educating masses of students in the secrets of mathematics.

Because the mathematicians are enshrined, everything beyond their own aristocracy constitutes the outside world. So, when Alfred Adler speaks of the mathematicians "lacking the imagination to appreciate the scope and sophistication of the outside world," he is referring to their attitude toward everything outside their own circle. In particular, the mathematicians consider the administration of their own university to be almost completely remote from their own existence. They view deans as failed professors and they see the complex layers of academic administration as a sanctuary for drones and incompetents who serve only to siphon away resources which should go directly to professors' salaries. The fifth question put to Timothy O'Meara about the "dumb provosts and deans" was not meant in jest.

Mathematicians particularly dislike serving on university committees and they hold administrators responsible for the existence of these committees. Professor Deep may resent having to interrupt his research to teach calculus but he is outraged at the notion that he should set aside his mathematics in order to spend an hour sitting around a table with administrators and outside faculty discussing student graduation requirements or some other obscure curricular matter. Deep believes committees exist only because administrators exist. If the administration vanished tomorrow, he believes, so would the committees and, with them, all the silly academic issues with which they deal.

Mathematicians have very few meetings of their own because they can't do mathematics at a meeting. When they do hold one of their infrequent departmental meetings nothing much happens—except loud and irrational talk. Alfred Adler wrote: "For example, departmental and mathematics-society meetings are occupied mainly with talk—aimless and pedantic talk, billowing with Latinisms. Little of substance is ever accomplished, or even intended."[22]

You could, however, call the mathematicians together for the purpose of voting on the question:

Resolved: All administrators other than departmental chairmen and departmental support staff are hereby eliminated.

All the mathematicians would show up for this one. The resolution would be carried by unanimous vote without the bother of discussion or debate. They would show up, call the meeting to order, call the question, and all vote "aye" within a matter of minutes. Professor Deep would leave his office, walk down the hall to the meeting, vote away the entire administration, and be back to his research faster than you can learn to spell "separable Lebesque structurable coalescences."

And—God help us—had the mathematicians the power to see to it, the administrators would disappear before Deep got back down the hall. As time passed, litter would pile up in the corridors, the grass would grow tall outside, and one by one the phones would fail. Other faculty—more attuned to the "outside world"—would notice the campus crumbling away around them. They would become aware of the loss of services: research grant applications not getting processed, applications for next year's admission piling up on a table somewhere, next semester's teaching schedule failing to appear. Soon, these outside faculty would notice something else: the absence of the limousines and helicopters that used to come and go bringing the well-heeled campus visitors to tour the facilities on the arm of the president and to leave behind generous contributions toward the university's building fund.

But the mathematicians would notice nothing. Deep would go on writing mathematics as before. When the lights failed because no one paid the electric bill, Deep would work by candlelight. When they came to take away his telephone, he would be inconvenienced only by having his research interrupted. The university can fall down like Jericho, but so long as the pencils and the yellow paper hold out, Deep's work goes on. At least a semester would pass before he even noticed he was no longer being paid.

I am not entirely unsympathetic with the mathematicians' view of academic administration except that they go too far. (Adler says that the positions mathematicians assume in bureaucratic affairs are "extreme and speculative.") To be sure, academic administrations have grown far too large and have become far too complex to respond efficiently and appropriately to the real needs of faculty and of students. I would like to see—as would the mathematicians—leaner and more academically oriented administrative organizations. But that is another story. All I am pointing out here—without making value judgments—is the inconsistency of the mathematicians' position toward the importance of the federal government as contrasted with their notion of proper university government: the former position being extremely liberal and the latter conservative.

You can see their conservativeness shining through in the responses to the challenges to clean up the calculus instruction mess. In conferences such as the one called *Calculus for a New Century,* you hear the mathematicians say repeatedly: "If there is to be meaningful change, it must come from *within* the mathematical community; it cannot be imposed from the outside." In the published report of the *New Century* conference, the mathematician Ronald C. Douglas wrote: ". . . change certainly cannot be dictated from the top down."[23] In the same report, Gina Bari Kolata, a science writer, stated: "Changes cannot be imposed on people and they have to occur gradually."[24] And, in his review of the conference report called *Toward a Lean and Lively Calculus,* the mathematician Richard W. Hamming wrote: "Reform of the calculus course is necessary but I believe it must come from within the mathematical community and cannot be imposed from without."[25]

Indeed. So believe all mathematicians. And when you place this belief alongside of a remark by Mathematical Association of America president Leonard Gillman you see clearly the extreme academic conservativeness of American mathematicians. Gillman wrote in spring 1988: "The calculus scene has been execrable for many years, and given the inertia of our profession is quite capable of continuing that way for many more."[26]

In other words, the calculus situation is loathsome but changes cannot be *imposed* on the mathematicians. We must wait until they have a change of heart. We may wait for "many years."

If you want to see an example of exaggerated behavior, I suggest that you sit with the mathematicians at lunch. Somewhere between the soup and the final cup of tea, say to them in your softest voice:

"Of course I believe the absence of minorities in the medical profession to be loathsome. But you cannot change the situation by imposing quotas on medical schools from the outside. You have to be patient and wait for the medical school admission people to have a change of heart."

You should be moving when you say it because, although physical violence is not part of the mathematicians' style, violence of expression is. You will have stirred them into a fit of fury. And it will be a storm you cannot shout down. Better to be moving and get in out of the wind.

You can also see the conservative nature of mathematicians reflected in the characteristics which people like Morris Kline and Timothy O'Meara assign to them. Words and phrases like "clannish, elitist, snobbish and highly individualistic" or "self-contained, inward-looking, cloistered" are terms normally not associated with political liberals. Look at the following three pieces of criticism which are entirely consistent with the description of mathematicians which has come forward from Kline, O'Meara, and others:

1. The *mathematicians'* weakness is not radicalism or extremism but parochialism. The ordinary *mathematician* looks within himself and purrs.
2. *Mathematicians* simply do not take much interest in the world around them.
3. The narrowness of America's *mathematicians* is a mystery.
4. The result is a *mathematics* composed of *mathematicians* who do not integrate their narrow values into the broad range of human experience.

The sentences could easily have been written by
Morris Kline or Timothy O'Meara or even by me. Actually
I lifted them from a March 1987 article published in the
Wall Street Journal. The article was titled "A Conservative
Crack-up?" and was written by R. Emmett Tyrrell, Jr.[27]
The article's theme was that the Reagan administration
had faltered because of "the failure of Ronald Reagan's
conservatives." Tyrrell saw this failure as deriving naturally from the static and unadaptable characteristics of conservatives. If, in the above four statements, you replace
"mathematicians" by "conservatives" and "mathematics"
by "conservativism" you have four sentences taken directly from Tyrrell's article. I find it fascinating that
Tyrrell's criticism of political conservatives applies word
for word to the academic behavior of the collection of politically liberal mathematicians. In a certain sense, the mathematicians join together the two ends of the political
spectrum.

There is a name for this joining. Look once more at the
sketch of the real number line, shown way back in Figure 2.
One of the characteristics of the real line is that it is un-
bounded. You can select any real number to the right of the
origin, no matter how far away, and then find another real
number farther to the right. Similarly, the real line is un-
bounded to the left, in the negative direction. In calculus,
we speak of the behavior of functions $f(x)$ as x tends to
"positive infinity," meaning "the limit of $f(x)$ as x becomes
arbitrarily large in the positive sense." Similarly, we can
speak of the behavior of $f(x)$ as x tends to "negative infinity." But these notions of positive and negative infinity are
simply part of the concept of limits and there are no positions on the real line which represent either. The real line
extends indefinitely in each direction.

You can, however, replace the real line—in a one-to-one manner—by something which is bounded. Namely, the real line can be mapped onto a circle in such a way that each point on the line corresponds to exactly one point on the circle and—with one exception—each point on the circle corresponds to exactly one point on the line. The construction of this correspondence is shown in Figure 15.

What you do is place a circle of radius ½ on the real line in such a way that it is tangent to the line at the origin and so that the point with planer coordinates (0, 1) represents the circle's "north pole." Select an arbitrary point x on the real line. Draw the line from x to the north pole of the circle. Let p denote the point where this line intersects the circle. The point p represents the image of x under our mapping. Conversely, select any point q on the circle, other than the north pole. Now draw the line from the north pole through the point q. Extend this line until it intersects the real line. Call the point of intersection w. Then w denotes the image of q under the mapping.

Clearly each point on the real line gets assigned, via this process, to exactly one point on the circle. Each point on the circle—other than the north pole—is associated with exactly one point on the real line. Points near the origin on the line go to points near the circle's south pole.

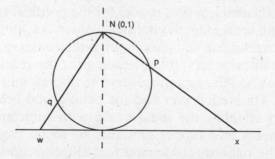

FIGURE 15. Circle projected onto a line.

Points near the north pole on the circle go to points far, far out on the line.

Thus, this "stereographic projection" replaces, in a one-to-one manner, the real line by something that is bounded. Namely, the real line is replaced by a circle with the north pole removed. If we wish, we can deal with the north pole by adjoining to the real line a point to which it corresponds. Since points close to the north pole are sent, under our mapping, to points far away from the origin on the real line, it is natural to call this adjoined point "the point at infinity."

When we add the point at infinity to the real line and assign it in this manner to the north pole, the mapping replaces the extended real line with something that is not only bounded but closed. Precisely, the real line plus the point at infinity gets replaced by the closed circle with center at $(0, \frac{1}{2})$ and radius $\frac{1}{2}$. Mathematicians call geometric objects, which are both bounded and closed, *compact*. The stereographic projection process we have just described is referred to as the "one-point compactification of the real line." The point at infinity is also referred to as "the ideal point."

When people speak of conservatives as being to the right and of liberals being to the left, they essentially have in mind the real line as a model for the political spectrum. The origin represents the political center. As you go to the right along the line you pass through the conservatives and then the ultraconservatives, heading for the reactionaries. When you go left, you encounter the liberals and then the radicals. On the left you find the people who believe that they can establish the ideal world with sufficiently large governments and proper amounts of social engineering. Out to the right are those who find all government an anathema, a curse imposed on them by people of less stature and ability.

Interestingly, you find the mathematicians at both ends. In terms of governmental politics they live far to the left. But, in terms of the way they deal with the political and bureaucratic affairs of their own academic environment, they stand to the right, far beyond all other subsets of the academy.

This characteristic may be unique to mathematicians. If so, then they become the proper choice for the "point at infinity" in the circle model of the political spectrum. The mathematicians form the north pole of the circle and represent the point that "compactifies" the line. They uniquely join together the political far right with the far left. The mathematicians—in this model—stand as the one-point compactification of the political spectrum.

I believe the model has some validity and that it identifies something unique about mathematicians other than their membership in the aristocracy of elegance. The model gives the position of the aristocracy on the line representing the political spectrum or on the circle to which the line corresponds point by point. The model tells you that the mathematicians occupy a unique point in either setting: on the circle they live at the north pole, on the line they are cloistered at the point at infinity. The aristocracy of elegance lives at the ideal point.

Good. It is appropriate that there exists something "ideal" about mathematicians since there is much about mathematics to which this adjective applies. In fact, you can construct the line–circle model for the spectrum of academic disciplines exactly as was done for the political spectrum. In this model, it is natural to place the sciences, and the disciplines which can be made "sciencelike" to the right. One way to order these disciplines is to place them on the positive real axis in the way in which they appear in the hierarchy of L. T. More. This means that sociology lies just to the right of the origin. Beyond sociology is history, and beyond history, you find psychology. Then, in order as you walk to the right on the line, you come to biology, chemistry, and physics. The other sciences may be placed on the line according to what More calls "their development from the accumulation of statistics of phenomena to the generalization of these phenomena in comprehensive and rigorous laws."[28]

In the L. T. More ordering, pure mathematics must be "greater than" any of the sciences or the near sciences, since *position in the order depends exactly on the degree of mathematical development of the subject.* Consequently, mathematics must be placed to the right of all other disciplines in the line model.

The other academic disciplines—in particular, the arts and humanities—are placed on the line to the left of the origin. The order in which these disciplines are set down is less clear than the ordering of the sciences in the other direction. G. T. More did not provide—as far as I know—an analogous hierarchy for the nonsciences. One way to obtain, at least, an intuitive listing would be to place the nonsciences in *aesthetic* order. That is, you place the "ugliest" nonscience closest to the origin and the more "beautiful" further out to the left.

Obviously, this method of ordering the nonsciences has more conceptual than practical value. In any particular case, an attempt to place one discipline to the left of another is sure to encounter difficulty. For example, I find it clear that, as an academic discipline, English literature possesses far more aesthetic value than does, say, the study of educational administration. Consequently, I would place

English literature far to the left of educational administration in this aesthetic ordering. But it would be difficult—and pointless—for me to try to write down specific criteria by which I made this judgment. And it would be sure to start an argument.

But the aesthetic ordering of the nonsciences is theoretically possible. In this ordering, mathematics is far out. Remember Bertrand Russell's remark: "Mathematics, rightly viewed, possesses not only truth, but supreme beauty . . . sublimely pure . . . capable of a stern perfection as only the greatest art can show."[29]

Yes. "Supreme beauty," "perfection," "greatest art." If you believe Russell—and I truly believe him—then you must agree that, in the aesthetic ordering, mathematics lies beyond all the nonsciences.

Thus, we find the discipline of mathematics occupying a unique position in the academic spectrum as do the mathematicians in the political spectrum. Mathematics—on the line—sits to the right of all the sciences and to the left of all the other disciplines. To make sense of this paradox, we replace the line in a one-to-one manner by the circle. Mathematics then, on the circle, lives at the north pole. On the extended real line, mathematics sits—ideally and uniquely—at the point at infinity.

We need next to examine this "uniqueness" of mathematics. In our model, mathematics joins together the sciences and the nonsciences. But this joining is only conceptual. We need to learn whether or not the unique position of mathematics in our combination of scientific and aesthetic orderings has any practical value.

CHAPTER 7

The Two Cultures

Decades have passed since C. P. Snow gave his famous Rede Lecture at Cambridge in 1959. In the lecture, Lord Snow said[1]:

> I believe the intellectual life of the whole of western society is increasingly being split into two polar groups. . . . Literary intellectuals at one pole—at the other scientists, and as the most representative, the physical scientists. Between the two a gulf of mutual incomprehension—sometimes (particularly among the young) hostility and dislike, but most of all lack of understanding.

In the intervening years, we have come to think of these two groups as the "humanists" and the "scientists" and we refer to them, using Snow's phrase, as *the two cultures.*

Lord Snow's lecture has been endlessly analyzed. Following the lecture a flood of articles and letters appeared almost immediately and references to it continually show up in the margins of contemporary works dealing with al-

most any aspect of educational philosophy. Taken together, these analyses are equivocal. Snow has been equally
blamed and praised.

But three things are absolutely clear: The two groups
exist, the gulf between them is real, and the phrase "the two
cultures" has become part of the English language. We all
know the phrase. And all you need do to convince yourself
of the existence of the two groups and the separating gulf is
simply visit any major college campus and *observe*. On any
campus, you will find that the scientists have little understanding of the work of the humanists and the humanists
understand science not at all. What is more significant, you
will discover that neither group appreciates the work of the
other. The scientists believe the literary intellectuals to be
shallow and woolly minded. The literary intellectuals think
the scientists narrow and barely literate. And you will find
the separating gulf containing more "hostility and dislike"
than "incomprehension."

You will find these things with probability one. The
certain finding of the two groups is not a matter for analysis, or psychological deduction, or even debate. It is a fact
of concrete experience. Just go look.

Much of the criticism of Snow's concept has to do with
whether or not these separate groups are actually cultures
and whether there are, in fact, *two* of them. Thus, there
appeared a flurry of articles examining Snow's notions of
"scientists" and "literary intellectuals" and contrasting
them with various interpretations of the idea of "cultures."
Snow responded to each criticism. In 1964, he asserted that
there were two acceptable meanings of "culture"[2]:

> First, "culture" has the sense of the dictionary definition, "in-
> tellectual development, development of the mind." . . . The
> word "culture" has a second and technical meaning, which I
> pointed out explicitly in the original lecture. It is used by

anthropologists to denote a group of persons living in the same environment, linked by common habits, common assumptions, a common way of life.

Lord Snow argued—effectively, I think—that either meaning was applicable to his theme and that the word "culture" "is still appropriate and carries its proper meaning to sensible persons."[3]

Regarding the second criticism—that it was inappropriate to speak of *two* cultures—Snow himself had "qualifying doubts." In the Rede lecture, he said[4]:

> The number 2 is a very dangerous number: that is why the dialectic is a dangerous process. Attempts to divide anything into two ought to be regarded with much suspicion. I have thought a long time about going in for further refinements: but in the end I have decided against. I was searching for something a little more than a dashing metaphor, a good deal less than a cultural map: and for these purposes the two cultures is about right, and subtilising any more would bring more disadvantages that it's worth.

These criticisms sometimes blend together, which makes it hard to tell exactly what part of Snow's arguments are being refuted. Sometimes all you can discern is that the critic dislikes what Snow had to say. I am not surprised that this more or less vague criticism often comes from the humanist side of the gulf. Jacques Barzun, for example, writes in his book *A Stroll with William James* (my italics)[5]:

> In describing the haphazard interchange between art and science within *what is obviously one culture not two,* the word science has been used as if it was always one thing from the time of its public triumph to the present.

It's hard to tell what Barzun means here. The "one culture" part is clear but which culture remains is not. Yet, the idea of "haphazard interchange" seems consistent with Snow's notion of gulf which, if it exists, must separate at least two groups. Moreover, Barzun seems to suggest that science changes too much to be clearly defined. And this point is particularly curious since, three pages later, he writes (my italics): "So we may say that science is *one domain, its boundaries fixed* by its assumptions."[6] This concept of fixed boundaries seems inconsistent with change: even laymen speak of scientific change in terms of pushing back boundaries.

And Allan Bloom in *The Closing of the American Mind* says of Snow's ideas[7]:

> Some may consider this labeling trivial, akin to C. P. Snow's calling science a "culture." Science may appear creative only because we forget what creativity really means and take it to be cleverness at proposing hypotheses, finding proofs or inventing experiments. . . . The project was inspired by C. P. Snow's silly conceits about "the two cultures." . . . For the scientist the humanities are recreation (often deeply respected by him for he sees that more is needed than what he offers, but is puzzled about where to find it), and for the humanist the natural sciences are at best indifferent, at worst alien and hostile.

It's difficult—for me—to make sense out of Bloom's remarks. On the one hand, he attacks the notion of science as a culture. But his supporting statement is about "creativity," and not culture. And the part about "hypotheses" and "proofs" sounds more like mathematics than science. Then he derides the notion of "two cultures" but, with the next breath, describes the alien and hostile view the humanist has of the sciences—something which sounds very much like Snow's "gulf." One thing is clear: Bloom lacks sympathy for Snow's ideas.

As for Jacques Barzun, you wonder why a man who ardently admires the pragmatism of William James does not simply take a campus stroll and *observe* that the arts faculty and the scientific faculty live in two distinct, nonintersecting, disjointed worlds that in no way can be considered "one culture." And it is puzzling how Allan Bloom, who writes so eloquently and persuasively of the need for educational reform, can be so vitriolic toward an idea for which Snow took little personal credit and which—if used properly—can itself be a force for the reworking of the educational apparatus.

I am afraid that, through their uninformed notions of science as something considerably less than it is, Barzun and Bloom have identified themselves as confirming examples of the validity of Snow's concept. They sound to me as the examples Snow might have in mind when he described the nonscientific "other side" of the gulf[8]:

> But what of the other side? They are impoverished too—perhaps more seriously, because they are vainer about it. They still like to pretend that the traditional culture is the whole of "culture" as though the natural order didn't exist. As though the exploration of the natural order was of no interest either in its own value or its consequences. As though the scientific edifice of the physical world was not, in its intellectual depth, complexity and articulation, the most beautiful and wonderful collective work of the mind of man. Yet most non-scientists have no conception of that edifice at all. Even if they want to have it, they can't. It is rather as though, over an immense range of intellectual experience, a whole group was tone-deaf. Except that this tone-deafness doesn't come by nature, but by training, or rather the absence of training. As with the tone deaf, they don't know what they are missing.

I hope I am wrong about Barzun and Bloom. I respect each man's work. But I'm not sure I should pay them much mind when they talk about science.

I know Snow is not wrong when he speaks of two intellectual groups and of the gulf between them. And I know he is absolutely correct in the above description of the "other side" and its "tone-deafness over an immense range of intellectual experience." Snow does not mean to imply here

—and neither do I—that the scientific side is unflawed. Before he wrote the above passage, Snow described the scientists' *impoverishment.* He pointed, in particular, to the absence of art—with the exception of music—in the scientist's life and to the dearth "of the books which, to most literary persons are bread and butter." Snow says that—of novels, history, poetry, and plays—the scientist reads "almost nothing at all."[9]

Moreover, many scientists have no respect for those whose world turns around books and reading. Once, back when I was associate dean of my university's college of arts and sciences, I served for a year on a committee which was concerned with the advancement of the sciences. One of the committee members was a distinguished bioscientist whose own world was reduced to what he could see through a microscope when he looked at the insides of catfish. He knew more than anyone living about the causes and the development of malignant tumors in marine creatures and his publication record was longer than any fish that ever got away. But what he knew about literature would fit inside your watch. His name was Cell and he went everywhere wearing his white laboratory coat and carrying a bulging folder marked "research." He sat next to me at the committee sessions. Cell considered me to be an advocate for the notion of liberal education which included a strong dose of literature and reading. Consequently, he considered me a challenge. To Cell, I was someone who needed reforming.

One of his favorite premeeting pastimes was to explain and reexplain to me the difference in the work-ethic notions of the scientists and the nonscientists. The nonscientists, he believed, simply did not work as hard as the scientists. He supported his argument by pointing out to me that, on any evening up until midnight, I would find the biology building blazing with light, indicating the fervent work going on in the laboratories inside. But the humanities building, he told me, was every evening dark as a cemetery. "Those guys," he asserted, "have nothing to do." By "those guys" Cell meant every humanist, everywhere.

"They work in their offices during the day," I told him. "At night, they work at home or in the library. They don't have to be in a laboratory."

"Baloney," he said. "They have nothing to do. Day or night."

One day Cell ventured outside his laboratory. Wearing his white jacket and carrying his research folder, he wandered down a corridor in the humanities building. He walked past the open office door of Professor Prose, the university's distinguished authority on the modern American novel. What Cell saw so excited him that he rushed over to my office to tell me about it.

"*I've* been working in my laboratory all morning," he said as he thrust his open research folder toward me with both hands. I got a glimpse of a black-and-white photograph taken through a microscope. It looked like a piece of honeycomb that had been licked by a bear. "*I've* been doing research," he told me.

"Good for you," I said.

He put the folder down on the corner of my desk. "What do you pay that guy Prose over in English?" he asked.

"No professor works for me," I said. "And you know I cannot discuss particular university salaries."

"You probably pay him a lot," Cell said. "Way too much."

"He's a distinguished and a senior professor," I said.

"Do you know what the son of a bitch is doing with *his* time?"

"I haven't a clue," I said.

"At this moment the bastard has his feet on his desk and he's reading a novel."

"What novel?" I asked.

"I don't know. Something with two A's in the title. Like Abalone, Abalone."

Professor Cell, as I said, deals with things that live underwater.

"Could it have been *Absalom, Absalom?*"

"That's it," he said.

"Do you know what Prose teaches?"

"Touchy, touchy, feely, feely," Cell said.

"Prose teaches a graduate course on Faulkner, Hemingway, and Joyce."

Cell only shrugged.

He took the research folder off my desk and clutched it to his chest, next to his heart.

"I told you those guys have nothing to do," he said.

The scientists—some of them—are badly impoverished.

However, the impoverishments of the two groups are of different orders of magnitude. Had they the inclination or the motivation, the scientists could come—however haltingly—to the humanistic or the literary world. Shakespeare is not out of a physicist's reach. Whether a physicist does, or does not, frequently and seriously turn to the Bard's sonnets depends only on his personal choice. However, on the other side, the situation is immensely different. Nonscientists have no conception of the "edifice" of science. And—as Snow put it—"even if they want to have it they can't."

The scientists play on even in the nighttime. But the others cannot, like Jessica in the *The Merchant of Venice,* mark the music. To the music of science, these others are tone-deaf.

Twice in the 1959 Rede Lecture, C. P. Snow said: "I

intend something serious." So do I. I intend to identify the cause of the humanists' tone deafness to the music of the sciences. And I will argue that this deafness comes not from something the humanists bear like a virus but rather from something they lack. The scientists possess an agent which the humanists do not have. The presence of this agent enables the scientists to understand nature through the creation of metaphors for reality and the building of theories which explain and extend the metaphors. The agent gives scientists power of the mind as the Lady in the Lake gave Arthur power of the sword. With it the scientists replace the confused and fragmented experiences of reality with ordered and meaningful abstractions. Using the agent, the scientists paint the broken shards of nature which they experience into an unending collection of pictures, each more impressionistic than anything imagined by Cézanne.

The absence of the agent keeps the humanists away from all this as the absence of a ladder kept the fox away from the grapes. The goal of science is to arrange the collection of impressionist pictures into Snow's edifice as you arrange pieces of broken glass into a great mirror. And all of this—the abstractions, the metaphor, and the impressions —must be considered as art of the highest order. But, lacking the agent, the humanists stand outside this activity. The absence of the agent keeps all of the high art of science beyond the conception of the humanists. And its absence causes the humanist to despise science exactly as Acsop's fox despised the grapes. "Science is sour," say the humanists. The missing agent, of course, is mathematics.

TYPE M

In spring 1979 I drove west to give two invited talks at one of the state universities in Pennsylvania. The talks

went well and in the evening I was pleasantly wined and
dined by faculty colleagues. After dinner I strolled back to
my hotel room feeling fine in the soft spring night. Next day
I retraced my journey and drove back east. About noon-
time, I crossed the Susquehanna River just outside Harrisburg. The day was clear, the sun through the open window
sat easily on my arm. Pinchas Zukerman played Beethoven
on the public radio station I'd located. The sun gleamed on
the water. All was right with the world.

Then, just then, as Zukerman danced his bow across
the strings in the fast part of the Beethoven Violin Concerto, the music stopped. An announcer interrupted and,
trying hard not to sound excited, updated the day's major
breaking news story: the evacuation plans following the
near meltdown at the Three Mile Island nuclear power gen-
erating plant. Three Mile Island, he said, is located on the
Susquehanna River near Harrisburg.

It was not like looking into the barrel of a loaded pistol.
Time did not stand still. But things changed. Beethoven
went out of my head. Yesterday's successes and the evening's wine and conversation became unimportant. Spar-
kle vanished from the water. The river turned dark as a
spider.

I looked upstream to my left and then downstream to
my right. The forbidding and familiar fattened cylinder
shapes of the cooling towers could not be seen in either
direction. But I was not relieved. Helpless, I could only roll
up the window and drive faster.

Two weeks later, I sat at lunch in the faculty dining
room. By random choice my luncheon companions were
two assistant professors: a young woman from sociology
and a young man from physics. We talked about Three
Mile Island. I told my story of the newsbreak and the coinci-
dence of first hearing about the accident when I was exactly

over the Susquehanna River. I did not tell them about feeling helpless. Grown-up men of my time and place do not acknowledge helplessness. Not out loud anyway.

When I finished, the sociologist spoke. She referred to the accident as a "disaster" using the term still current in the media even though no injuries had been identified. She spoke of past nuclear disasters and the potential for future disasters. She talked of white men dropping nuclear bombs on yellow people. She mentioned Robert Oppenheimer's famous remark: "Now I am become death, the destroyer of worlds." She talked about capitalist greed and the failed ethic that places concern for electrical power above concern for people. And she spoke of a dark future filled with birth defects and lingering illnesses resulting from the health hazards associated with nuclear power.

She was concerned and passionate. She talked without pause as her fruit salad warmed and her tea cooled. I nodded noncommittingly from time to time and ate my tuna sandwich. The physicist sat listening and motionless. When she finished he said to her quietly:

"You don't know what you are talking about."

She bristled. "Don't bother telling me that no one was hurt so it isn't a disaster. All we know is that no one has yet been identified as being hurt. We don't know what nuclear poison the residents are carrying around inside them."

"That's not my point," the physicist said even more quietly.

"Then what is?"

He took a small notebook from his jacket pocket and then an old-fashioned fountain pen. A real one with ink that comes from a bottle. He took his time unscrewing the cap. He wrote something on a blank page and passed it across to her. She held it so I could see. He had written a single equation:

$$\frac{dy}{dt} = ky.$$

"So what?" she said.

"Do you know that this means?" he said, pointing to the equation.

"I am not a mathematician," she said.

"Neither am I," he said. "You don't have to be one to understand this equation. We teach it to freshmen. Ask the dean."

She looked across at me and I nodded again.

This time the young physicist's voice had a slight edge: "It's the differential equation which describes nuclear decay. When you solve it you get an explicit expression for the decay. When you manipulate the solution in an elementary manner you can determine the half-life of the nuclear substance. You can't talk about future health problems unless you understand these things. And you don't. All you have to say is air. Nothing but air."

He paused and sipped his tea. She looked across at me again. Lost and helpless.

"I'd write the solution of the equation for you," he said, "but you wouldn't understand that either."

Without a word she rose and left the table and the room. I finished the last bite of my sandwich and walked out behind her. The young physicist remained alone with his lunch, smug and satisfied.

In the evening I related the lunch table events to my wife.

"It was no contest," I said. "He's type M. She is type N."

"What's type M?"

"The M stands for mathematics," I said. "He has some facility with mathematics. Such people are of type M."

"Do you mean something inherent? Like blood type?"

"Absolutely not," I said. "You become type M by *learning* some mathematics. You aren't born that way. What I mean by 'facility' is a certain level of skill and knowledge. Anyone can become type M, it requires only study and practice."

"What does type N mean?"

"The N just stands for 'not M.' People who are not type M are of type N. Type N people have no real mathematical skill."

"Interesting," she said.

"It's more than that," I said. "It is *fundamental*. People of type N cannot argue science or technology with people of type M."

"Why?"

"Because they always lose."

"Are you sure?"

"Yes," I said. "They lose even when they are right."

C. P. Snow warned that any attempt to divide a set of objects into two parts should be regarded "with much suspicion." Nevertheless, it can always be done. All you need do is look at the objects and identify any particular property which some but not all of the objects possess. The set then falls naturally into two parts: those objects with the property and those without.

Consider, for example, the collection of professors who teach at First College of Liberal Arts. A few of these professors are over six feet tall; tne remainder are six feet or less in height. Thus, the set of First professors can be separated naturally and precisely into two groups: those over six feet tall and those who are not.

Similarly, you could divide the set of professors into

the group consisting of those who have blue eyes and those who do not. Or into the subset of those who can recite from memory Shakespeare's sonnet number 18 and the subset of those who cannot. In each case, of course, the division is only theoretically precise. Placing a particular professor into his proper group will depend on the accuracy with which his height can be measured, his eye coloring determined, or the manner in which his knowledge of the eighteenth sonnet is evaluated. But, in theory, the separation into two groups is straightforward and routine. Just identify a property and define one group to be those objects which have it. The other objects live, by definition, in the other group.

An advantage of this simple method of separation of a set of objects comes from the obvious fact that each object falls into one group or the other. A particular First College professor's height exceeds six feet or it does not, his eyes are blue or else of some other color, he can recite the eighteenth sonnet or he cannot. Once you agree on a method of height measurement, eye-color determination, or verse recitation you can determine into which group any professor belongs. And no professor is left out. Each falls into one group or the other.

One of C. P. Snow's problems was that his two divisions of "the intellectual life of the whole of western society" omitted some people. Snow's two groups consisted of scientists on the one hand and the literary intellectuals on the other. Obviously not all intellectuals fall into one or the other of these categories. No matter how carefully Snow's groups are defined you will have trouble trying to place an economist, for example, into one group or the other. Clearly, there exist many people who lead intellectual lives but who are neither scientists nor literary intellectuals. Much of the criticism of Snow's ideas came, in fact, from the outside perception that he tried to force everyone into one or the other of his categories.

What is needed is a more clearly defined separation which serves the purposes of Snow's arguments, which more or less agrees with his division, and which leaves no one out. We need to redefine the "two cultures" so that no one is left out.

Such a division exists and the resulting two cultures exist. One culture consists of those people who have knowledge of mathematics. The other culture is composed of those who do not. The people who belong to the first culture are of type M. The members of the second culture are all of type N. A high wall separates these two cultures. The name of the wall is mathematics.

Obviously, one needs some reasonably clear standard of measurement of mathematics facility before the groups can be precisely determined (a reasonable standard would be to say that a person is of type M provided he has familiarity with the basic concepts and techniques of calculus). But, even without the determination of such a standard, two facts are clear:

- Any two people of type M recognize one another immediately.
- When restricted to those who lead "intellectual lives," the type M and the type N cultures closely approximate Snow's two cultures of the scientists and the literary intellectuals.

C. P. Snow, I expect, would place an electrical engineer and a theoretical physicist in his scientist category. And he would assign both a classicist and a historian to the literary intellectual category. So would I. But what the electrical engineer and the physicist have in common is something more fundamental and more basic than their commitment to a common methodology based on the scientific method of the testing of theory against careful experimentation. The theoretical physicist may, in fact, be far removed from any actual experimentation while the electrical engineer may be engaged in work so practical as to be mainly industrial development. But the engineer and the physicist do share a deep and underlying commitment to the principle that the language in which the book of nature is written is mathematics. They each can read—with differing levels of literacy—this language. When they discuss their work with one another, they will not hesitate to use mathematical terminology and symbolism. Each person is of type M. And the other knows it immediately.

On the other hand, the classicist and the historian are almost certain to lack mathematical facility. Neither is likely to have studied college mathematics even at the calculus level and neither of them will use mathematics either in his research or in his teaching. Their papers will be explanatory and descriptive. The papers may be important. But they will not be mathematical in the slightest. And, while this absence of facility with mathematics will cause them no concern so long as they talk only to each other or to others of type N, it will cause problems when they try to converse with people in the other culture exactly as it caused the sociologist at my lunch table. When an argument arises, the type M person, sooner or later, turns to mathematics. The type N person, then feeling illiterate, invariably withdraws.

A well-known example of such an encounter involves the eighteenth-century philosopher Denis Diderot and his contemporary, the great Swiss mathematician Leonhard Euler. You can find the incident related by Ian Stewart[10] and by Augustus De Morgan. De Morgan's version follows[11]:

> Diderot paid a visit to the Russian Court at the invitation of the Empress. He conversed very freely, and gave the younger members of the Court circle a good deal of lively atheism. The Empress was much amused but some of her councillors suggested that it might be desirable to check these expositions of doctrine. The Empress did not like to put a direct muzzle on her guest's tongue, so the following plot was contrived. Diderot was informed that a learned mathematician was in possession of an algebraical demonstration of the existence of God, and would give it him before all the Court, if he desired to hear it. Diderot gladly consented: though the name of the mathematician was not given, it was Euler. He advanced towards Diderot, and said gravely, and in a tone of perfect conviction: "Monsieur,
>
> $$\frac{(a + b^n)}{n} = x,$$
>
> therefore, God exists; répondez!" Diderot, to whom algebra was Hebrew, was embarrassed and disconcerted, while peals of laughter rose on all sides. He asked permission to return to France at once, which was granted.

What Euler said to Diderot was nonsense. But Diderot didn't know because he, a man of type N, was dealing with a person of type M. The argument ended—as all such arguments must end—with the type N person on the run.

When C. P. Snow spoke of two cultures he had in mind two divided sectors of western "intellectual life." The set of people he separated into two groups were people who worked in the academy or else those whose primary activities were of academic type—people such as writers, composers, or nonuniversity scientists. He was not concerned with society as a whole and his two cultures did not encompass those whose occupations were not, in some sense or another, primarily intellectual occupations. Snow did not extend his two-culture concept to shoe salesmen.

On the other hand, the type M and type N concepts divide all of civilization into two categories. Each human either has skill with mathematics or he does not. If he has, he is of type M. Otherwise, he falls into the type N group. But the type M and type N notions—like Snow's two cultures—are of interest mainly when they are applied to academic intellectuals. When this restriction is made, the type M and type N academics agree almost exactly with Snow's categories of "scientists" and "literary intellectuals."

While the type M and type N notions essentially coincide with Snow's idea of the two cultures, they actually give considerably more. First, these notions allow one to understand the extreme differences which can exist between researchers in the same field; researchers who—in Snow's scheme—should belong to the same culture but who obviously do not. An example of such a separation can be found in the field of international relations.

Scholars who work in the field of international relations deal with international politics, diplomatic decision making, foreign policy decisions, and the general movements of sovereign states relative to one another. At first glance, it seems as if the entire field falls outside of Snow's two categories of scientists and literary intellectuals. I have no doubt, however, that Snow and the supporters of his concepts would place these researchers, along with their close colleagues, the historians and the political scientists, in the literary intellectual category. But such a general classification would be simplistic because the researchers in this area are themselves deeply subdivided into groups which may be described by their differing analytic approaches to the subject matter of the field. These two approaches have been labeled by Hedley Bull the *classical approach* and the *scientific approach.* And Klaus Knorr and James N. Rosenau—who edited the book *Contending Approaches to International Politics,* in which Bull's description appears—describe the researchers using these approaches as *traditionalists* and *scientists,* respectively.

If you read Hedley Bull's essay you find that the division between these two groups is deep and bitter (at least up until the date of the book's publication). Hedley Bull, himself a traditionalist, says of the other side[12]:

> . . . the scientific approach has contributed and is likely to contribute very little to the theory of international relations, and in so far as it is intended to encroach upon and ultimately displace the classical approach, it is positively harmful. . . . they [the scientists] have done a great disservice to theory in this field by conceiving of it as the construction and manipulation of so-called "models."

On the other hand, Morton A. Kaplan writes in the same volume[13]:

> This [improper use of methods] is illustrated by the fact that so intelligent a student of politics as Hedley Bull, who openly recognizes the danger that he might be talking about discordant things, nonetheless falls into what I would call the trap of traditionalism: the use of over-particularization and unrelated generalization.

And Kaplan also writes: "The traditionalists are often quite intelligent and witty people. Why then do they make such gross mistakes?"[14]

And so it goes throughout the book: claim and counterclaim, criticism and countercriticism; a single field divided down the middle. But if you examine the division carefully, you will find that the terms "traditionalist" and "scientist" are not completely accurate. What distinguishes "scientists" is, once again, not any commitment to the scientific method nor to experimentation but simply that, in their research in the field of international relations, they make extensive use of mathematics and of mathematical methods. The "traditionalists," on the other hand, do not use mathematics in their work. The scientists in international relations are people of type M; the traditionalists are of type N. The division between them is deep and bitter exactly because type M people do not communicate professionally with people of type N. Type N people do not *want* to talk to people of type M mainly because they are intimidated by them. Communication in the other direction is by definition impossible because there can be no understanding.

Thus, the type M and type N notions—while being essentially consistent with Snow's two cultures concept— yield a finer division of intellectual society. They allow you to recognize and to understand the differences which often exist between researchers in the *same* field. But these notions provide one more thing, something far more valuable and fundamental than finesse and subtleties of separation. The notions show exactly what you must do to remove the separation, what must be done to bring the two cultures together.

In fact, the matter now can be described quite simply. The two cultures consist of the collection of individuals of type M and those of type N, the first type consisting of people who have facility with mathematics and the latter of those who do not. And we may think of "facility" as being defined by a familiarity with mathematics through the level of elementary calculus. The membership of the class of type M includes engineers and scientists and others who *use* mathematics in their work but who fail to experience it aesthetically. Type M people are too *close* to mathematics to find it aesthetically pleasing; they are "underdistanced" in the sense of aesthetic distance. Mathematics for them is too hot. They have come too close to it as Icarus came too close to the sun.

On the other side—overdistanced and far away—live the people of type N. These—the humanists, the poets, the artists—are out there somewhere in the dark. To them, the art object called mathematics might as well be a marble buried miles deep in polar ice. Mathematics is irrelevant.

These two groups are shown in Figure 16 where the aesthetic ring of mathematics has been produced once more. In between the two groups, alone in the eggshell-thin aesthetic ring of mathematics, live the mathematicians. Regal and wise, they pass the subject's great aesthetic pleasures from one to another, keeping them from the type M people who are their natural customers, and from the type N people with whom they interact not at all.

The aesthetic ring of mathematics defines and sepa-

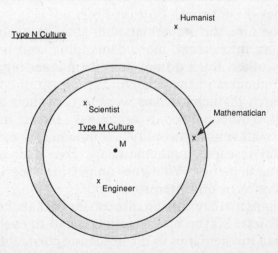

FIGURE 16. The aesthetic ring of mathematics and the two cultures.

rates the two cultures. And the size of the separation is determined exactly by the sum of the number of intellectuals who live beyond the outside boundary of the ring (the type Ns) and the number inside the inner boundary (the type Ms). If you want to bring these cultures together you have only one possible course of action—an action which would require nothing short of a revolution in the teaching of elementary and college-level mathematics. To bring the two cultures together you must widen the ring.

UNIQUENESS

Mathematicians, as far as I know, do not consider their subject *unique*. In my experience, at least, they do not talk about the uniqueness of mathematics. Even the college teachers of mathematics outside the research universities—people who teach mathematics as a profession and do talk about it—do not seem to speak of its uniqueness. Certainly they do not speak of this to me.

To be sure, the mathematicians consider their subject to be more intellectual, more demanding, and important than any other. But I do not hear them asserting the subject's uniqueness in the academic spectrum of courses. If the research mathematicians were to turn their attention seriously to other academic disciplines I expect the singleness of mathematics would be immediately evident to them. But research mathematicians give their attention only to mathematics. What goes on outside the aristocracy neither concerns nor interests them.

Perhaps mathematicians are too far back in the trees to see the forest. Maybe mathematicians fail to evaluate the position of mathematics in the academic curriculum as admirals often fail to ascertain the proper role of sea power in international diplomacy (an admiral's job is military work; diplomacy is something else). It may be that an informed, outside judgment is needed. One such judgment has been provided by Allen L. Hammond.

Hammond has considerable training and knowledge of mathematics but he is not a practicing mathematician. He stands outside the aristocracy of mathematicians and looks in. Possibly because of this, he sees mathematics with considerable clarity. In the article "Mathematics, Our Invisible Culture," he asks: "What is it in the nature of this unique field of knowledge, this unique human activity that renders it so remote and its practitioners so isolated from popular culture?"[15]

Hammond, an outsider, recognizes the uniqueness of mathematics. And, two sentences later, he comes almost to the crux of the matter. He writes[16]:

> For example, mathematics is nearly always described as a branch of science, the essence of pure reason. Beyond doubt mathematics has proved to be profoundly useful, perhaps even essential, to the modern edifice of science and its technological harvest. But mathematicians persist in talking about their field in terms of an art—beauty, elegance, simplicity—and draw analogies to painting, music. And many mathematicians would heatedly deny that their work is intended to be useful, that it is in any sense motivated by the prospect of practical application. A curious usefulness, an aesthetic principle of action; it is a dichotomy that will bear no little scrutiny in what is to come.

Yes. In five sentences Hammond almost has it all. Mathematics, to science, *is* "profoundly useful." And, yes, mathematicians think of their subject as an "art." Eugene P. Wigner, you will recall, spoke of the "unreasonable effectiveness" of mathematics in science. And John Polkinghorne, the Cambridge physicist, confirms both Hammond and Wigner with his remark, "Mathematics is the abstract key which turns the lock of the physical universe."[17]

Certainly, Hammond is correct when he asserts the existence of a curious and deep "aesthetic principle of action." Moreover, he properly notices the significance of the principle–action dichotomy. But he fails to follow through. While Hammond in his essay reports on this aesthetic principle and records some instances of mathematicians speaking to one another about it, he nowhere connects their failure to reveal it directly to the outside world with the deep division which exists between the type M and the type N cultures.

Because Hammond does not observe this connection, he comes up short. He fails to notice the enormous educational potential residing in the mathematicians' aesthetic principle. Something must be added to Hammond's otherwise fine summary. I call the addition the *uniqueness property of mathematics*:

> Among the undergraduate academic disciplines, mathematics stands *uniquely* alone as both art and science. Engineers and scientists appreciate mathematics because of its utility; without mathematics they cannot do their work. The others—the humanists, the literary intellectuals, all those of type N—can be touched by mathematics only if it is presented to them as the mathematicians themselves see the subject: as an elegant and creative art. The humanists cannot be engaged with mathematics by presenting it to them on the grounds of the subject's utility or its applicability. For this is the way mathematics has been offered to them already and this approach has only driven them away. And it has led to the very existence of the two cultures.

Mr. Hammond has "failed" only in not taking his thesis to its logical extension. The mathematicians' failure, however, is a failure of omission and one of great import. The mathematicians have failed to carry out what ought to be their natural mission, namely, the transmission to educated men and women outside the aristocracy their concept of the elegance and the art of mathematics. And this task can be performed by no others. Only the mathematicians understand the imaginative artistic experience revealed to those who truly contemplate the subject. The mathematicians alone understand the subject's aesthetic connection. They are intimate with Hammond's "aesthetic principle of action." But they have failed to pass even the principle's existence along to anyone else.

In *The Principles of Art,* R. G. Collingwood emphasizes the importance of the contact of the artist with his audience. He argues—effectively, I think—that this contact must be something more than mere communication from artist to audience. Collingwood asserts that, unless the contact becomes a kind of collaboration, the artist—and ultimately the art itself—will fail. He writes: "The kind of contact that is required is a collaborative contact in which the audience genuinely shares in the creative activity."[18]

And Collingwood tells us how the artist must stand with respect to his audience if he is to succeed[19]:

> Instead of conceiving himself as a mystagogue, leading his audience as far as it can follow along the dark and difficult paths of his own mind, he will conceive himself as his audience's spokesman, saying for it the things it wants to say but cannot say unaided.

It is self-evident that the mathematicians have failed —and failed miserably—to meet Collingwood's criteria. Collaborative contact between mathematicians inside the aristocracy and anyone outside simply does not exist. The members of the aristocracy have no notion that it is desirable—or even possible—to share their creative activities with an outside audience. And—as far as the humanist portion of the outside audience is concerned—the very existence of this creative activity is unknown.

Of all their shortcomings, the inability of mathematicians to bring the elegance and beauty of their high art to the attention of the humanists may be their greatest. For the humanists stand farthest from mathematics. Scientists come to mathematics because they have need of it, because it pays. Like it or not they must come to nature's language if they are to read its book.

But the humanists see no need for the mystagogy of mathematics so they stay away. Far outside the aesthetic ring, mathematics to them seems cold as a winter star. And because of their remoteness from the magical language of mathematics, there are things the humanists *cannot* say: things about nature and about modern technology about which they have strong emotions and important ideas but cannot express because the people who dominate the technical world are of type M and they, the humanists, are of type N.

These humanists are the natural audience of the mathematician: an audience for which he should be the spokesman, "saying for it the things it wants to say but cannot say unaided." But for this to work the mathematician must share a real sense of collaboration with the humanist audience. And this would require the mathematician to dramatically shift his notion of priorities and of value. As Collingwood put it[20]:

> This will involve no condescension on his part; it will mean that he takes it as his business to express not his own private emotions, irrespectively of whether anyone else feels them or not, but the emotions he shares with his audience.

Yes. But Collingwood did not have mathematicians in mind when he laid out this prescription for artist–audience collaboration. And I doubt that you can find half a dozen mathematicians who believe that the humanists compose an audience with whom they can share their mathematical aesthetic emotions. In their relation to the humanists, the mathematicians are Collingwoods' mystagogues—secretly passing mathematics' mysteries hand to hand, one to another, hidden somewhere inside a dark ring.

CHAPTER 8

Great Things

Bertrand Russell once came seriously to the question of why one should study mathematics. The answer he set down exceeded the narrowness of the question and provides an acceptable reason for why one should study anything. I see it, in fact, as a one-sentence statement of philosophy for the concept of a liberal education. Lord Russell writes: ". . . it is well to be reminded that not the mere fact of living is to be desired but the art of living in the contemplation of great things."[1]

Yes, the *art* of living in the contemplation of *great things*.

By great things, Russell meant painting, sculpture, literature, music, architecture, and all the rest that are normally considered under the general rubric of "art." But he meant more than these. "Great things" include *great ideas*. Russell's art of living requires one to be "fully alive to the beauty of contemplations."

So does mine. And one of the vastest areas of the world

of contemplative beauty is mathematics. This alone is sufficient reason for its study.

LOW ART

I did not begin with mathematics because of its beauty. Nor did anyone else I've known. Like all schoolchildren, I began the study of mathematics because I had no choice. What I was told—as were all the rest—was that mathematics was somehow good for me. The subject has—my teachers said—great practical value. "Just you wait," they said. "One day you'll see."

I wondered then why they did not tell me to "just endure." For that is what I did with mathematics for many years—I endured.

My teachers simply "told" mathematics to me. They gave me certain rules for manipulating numbers and symbols. My participation in the process required me only to practice the manipulation sufficiently so that I could reproduce certain demonstrations on homework papers and examinations. At no time, at any level prior to college, did my teachers indicate that there might exist some relation between mathematics and what I vaguely understood as liberal education. Mathematics, in my school days, was manipulation—only this and nothing more.

The rules they gave me seemed then to be complicated and I could only learn them by rote. I now know they are not and that the apparent complexity came only from the manner in which they were presented, and from the method with which I was supposed to fix them in my head —a method consisting only of unthinking, day-after-day repetition. Now, I also know my teachers presented mathematics as they did because this was the way it had been presented to them. They understood it no more than I.

Like many youngsters, I was puzzled by negative numbers (see p. 73). One day I said to my teacher: "I don't understand why -2 multiplied by -3 gives $+6$."

"You have," she answered sternly, "a very bad attitude toward mathematics. I've already told you that the product of two negative numbers is always positive."

In those days, mathematics was simply *told*. And telling is less than teaching. Nothing, I am afraid, would have changed, had my teachers known what Robert Henri, an influential American painter, said of "telling": "Low art is just telling things; as, There is the night. High art gives the feel of the night."[2]

My early teachers taught mathematics as low art. They gave no *feel* for the subject. Sadly, they had none to give.

From my experience with beginning college students and from my knowledge of their mathematical backgrounds, I conclude that little has changed. Students come to college with no feel for mathematics that can, in any way, be associated with art. In fact, they seem convinced that either no such sensation is possible, or else that it falls—like ultrasonic sound—outside the sensory range of ordinary mortals. They come to college—as once did I—after many years of sitting in classrooms and having mathematics told to them.

As I have tried to make clear, I do not believe that a feel for mathematics is innate. In fact, part of this book's thesis is that mathematics as art can be made accessible to a wide audience, certainly to those people who find other arts such as music, literature, and painting indispensable. The fact that mathematics presently lies outside the artistic range of most people is the fault of neither the audience nor of mathematics. What has gone wrong is the manner of presentation. How else can there exist a person who likes poetry and hates mathematics? Properly presented, they are much the same.

Students come to college after too many years of experiencing mathematics as low art. Unfortunately, many of them leave the same way.

VALUE

My early teachers chanted the notion of practical value like a litany. It was repeated at each level, in each course, from grade one through high school. By "value" they did not mean "worth" in the sense Bertrand Russell uses the term. Instead, my teachers intended something mundane. They meant to justify mathematics on the basis of its utility in the conduct of one's daily life.

There is nothing wrong with this except they went too far and claimed too much. Mathematics *is* useful in this sense. But, with this narrow connotation of "value," a little goes a long way. Counting change, measuring carpet, or balancing one's checkbook requires only the slimmest knowledge of mathematics. From early on, I wondered why such pedestrian activity required so much schooling.

The true value of mathematics lies outside commonplace activity. Essentially, the "value of mathematics" has two components. They are:

1. The value of mathematics as one of Russell's "great things," something necessary for life as "art" and not just "fact."
2. The value of mathematics as Polkinghorne's "abstract key which turns the lock of the physical universe."

As we have seen, the first of these concerns the intrinsic worth of mathematics itself as a creative and intellectual art. The second value, as we know, stems from mathematics' unreasonable effectiveness in explaining and predicting real-world, physical phenomena. To fully appreciate either of these values, one must seriously study mathematics up to some level.

My purpose here is not to present such mathematics. Rather, I have tried to describe mathematics in such a way that the existence of these values is made clear. In particular, I wanted to dispel forever the notion that the worth of mathematics is, somehow, paltry.

As for serious study, the exact level at which full appreciation of mathematics occurs is by no means obvious. In fact, if the word "full" is strictly interpreted, a clear-cut level probably does not exist. Nor, for example, would an exact literature level exist for full appreciation of the modern English novel.

However, for most of us, "significant mathematics level" means "mathematics study through calculus." You will recall that it was this level I had in mind when I discussed the characteristics of the type M and type N cultures. My rule of thumb then was that a person is of type M if his working mathematical knowledge includes a reasonable amount of calculus.

In this discussion of level, it might be helpful if you understand the peculiar role calculus plays in the mathematics curriculum. For most people, calculus represents the archetype of higher mathematics. It is the highest mathematics course to which they aspire. To most mathematicians, on the other hand, calculus stands—at best—as the first legitimate course in mathematics. All that stuff before calculus represents—to many mathematicians—only pre-mathematics training. It's merely preparation, as New Haven is preparation for Broadway.

Please believe that I am not patronizing. I am simply describing the situation as mathematicians see it. In truth, I do not share their view. I have taught calculus now for thirty years. Yet, each time I teach, I discover some aspect of the subject I did not know before. I consider calculus teaching to be the most significant part of my job, the most important work I do. I do not find it condescending.

To gain a fix on the value of calculus, you will need an intuitive notion of the subject. Think of calculus as a great arch of ideas fitted together like stones (see Figure 17).

The earliest and roughest ideas came from Archimedes of Syracuse around 200 B.C. and from Pierre Fermat in seventeenth-century France. Isaac Newton (1642–1727) in England and Gottfried Leibniz (1646–1716) in Germany

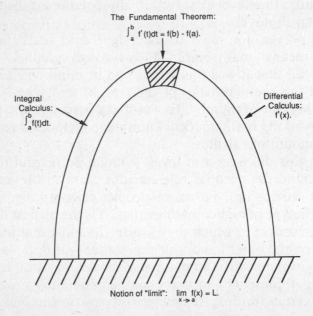

The Fundamental Theorem:
$$\int_a^b f'(t)dt = f(b) - f(a).$$

Integral Calculus:
$$\int_a^b f(t)dt.$$

Differential Calculus:
$$f'(x).$$

Notion of "limit": $\lim_{x \to a} f(x) = L.$

FIGURE 17. The calculus arch.

independently, and almost simultaneously, created, cut, and set the majority of the arch's stones.

The prongs of the arch have names: one is called *differential calculus* and the other *integral calculus.* These prongs represent the apparently dissimilar notions of the derivative of a function and the integral of a function. Much of the greatness of Newton and Leibniz derives from their completion of the arch by joining together the prongs. Newton and Leibniz found the keystone, the crown stone which binds together the prongs at the top. The keystone is called *the fundamental theorem of calculus.*

It is impossible to overestimate the value of this accomplishment. Above the great arch and supported by it rests all of mathematical analysis and the significant parts of physics and the other sciences which calculus sustains and explains. Mathematics and science *stand* on calculus as, in Florence, shops that sell you the finest silk, porcelain, and gold stand on the Pont Vecchio.

The dissimilar notions of derivative and integral, from which the prongs of the arch take their names, themselves came from something more fundamental. Like much of what is now abstract mathematics, these concepts arose from serious attempts to understand certain real-world phenomena. The derivative and the integral came, respectively, from the desire to answer the questions:

i. What is the area of a region bounded by a curved line?
ii. What is the instantaneous velocity of a moving particle?

The first question is the more ancient of the two and important contributions were made by the early Greek mathematicians, particularly by Archimedes. In particular,

Archimedes attacked the problem of finding the area of a circle. His method—the famous method of exhaustion—consisted of approximating the circular area ever more closely by triangles (whose area he knew), and then using an intuitive "limit argument" to deduce the correct area.

Archimedes' method was extremely powerful and, when applied to more general "curved areas," say, the area of a region bounded by something as irregular as the shoreline of a lake, it becomes the forerunner of integral calculus. In fact, the symbol in Figure 17 which looks like an "elongated S" is called an "integral sign" and represents a kind of "infinite sum" which is a direct generalization on the intuitive process which the lofty Archimedes devised 2,200 years ago.

The expression, $f'(x)$, which appears on the other prong of Figure 17, stands for "the derivative of f with respect to x." This new function, which is determined by another kind of limiting procedure, represents the rate of change of a given function $f(x)$ with respect to x. In the special case in which x represents time and $f(x)$ represents the distance a particle travels in that time, $f'(x)$ turns out to be the instantaneous velocity of the particle at time x.

The careful development of these concepts (which I have glossed over in fewer than 200 words) is not easy and involves, I think, reasonably sophisticated mathematics. Nevertheless, the material belongs to calculus and is taught —in one form or another—to freshmen at most leading universities.

On the face of it, there seems to be no reason why questions i and ii above (and the resulting notions of integral and derivative) should be related. What should the problem of determining, say, the surface area of Lake Tahoe have to do with the velocity of a falling coin, pitched from the top of the Empire State Building?

However, the problems *are* related and, consequently, so are the notions of integral and derivative. The great accomplishment of Newton and Leibniz was their (independent) demonstration of the exact nature of this relationship. These men proved, at a reasonable level of rigor, that the integral and derivative are related by the equation shown at the top of Figure 17. Briefly, this equation says that the integral and the derivative are "inverse" to one another in the sense that "the integral of the derivative of a function equals the function itself." When this phrase is stated precisely, the result becomes "the fundamental theorem of calculus," which stands without question as a paramount achievement of human thought.

The worth of these ideas is beyond price. Without them, there would be neither science, nor technology, nor understanding beyond simple observation of the physical world. Galileo says, "The book of nature is written in mathematics." In this statement, you could replace "mathematics" by "calculus" and not be far wrong.

The ideas of calculus are extendable to other, and higher, mathematical subjects (this is one of the reasons mathematicians consider calculus to be an entry-level course). The extension closest to my heart is the subject known as "complex analysis." Here, the methods of derivative and integral are generalized so that they apply to functions $f(z)$, where z is a complex number.

Calculus-type theorems, when extended to complex functions, become even more powerful and more lovely. One of these results, known as the Cauchy Integral Formula, is illustrated in Figure 18. Here, C is a curve in the complex plane, z a complex number which lives inside C, and ζ (zeta) a complex variable which lives on the curve. The theorem expresses the remarkable fact that—under a certain simple hypothesis—the value of the function at z,

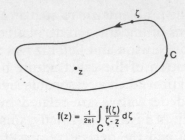

$$f(z) = \frac{1}{2\pi i} \int_C \frac{f(\zeta)}{\zeta - z} \, d\zeta$$

FIGURE 18. The Cauchy integral formula.

given by $f(z)$, can be determined by its values, $f(\zeta)$, on the curve C.

A real-world analogy of this result might go something like this: imagine a copy of next Sunday's *New York Times,* especially printed with type on only one side of each page. Spread the entire newspaper flat on the floor with the pages touching each other. (You'll need a huge floor; this one-sided edition is twice normal size. The playing surface of Madison Square Garden should suffice.) With a black marking pen, trace a closed curve around the exterior edges of the spread newspaper. Draw the curve so that it just touches each word at the paper's outer edge. Now go carefully around the closed curve, like the ζ in Figure 18, reading each word which touches the curve. Then invoke the Cauchy Integral Formula. Like magic, the theorem gives you every word in the entire paper.

I realize that I am merely *asserting* the aesthetic value of the Cauchy Integral Formula. The result lies too far beyond the scope of this book for me to do otherwise. It's not possible here to examine the proof of the theorem or to discuss the mathematical ramifications of its conclusion. Nevertheless, the aesthetic value exists and—at the appropriate mathematical level—it becomes clear that the result satisfies both the principle of minimum completeness and the principle of maximum applicability. If you come to it properly prepared, you will react as did knowledgeable audience members to the 1913 Paris debut of Stravinsky's *Le Sacre du Printemps*. It will take your breath away.

To be sure, the Cauchy Integral Formula lies far beyond ordinary calculus. But the fundamental ideas which lead to it are calculus ideas. In fact, it is possible to trace essentially all of calculus—and the subjects like complex analysis which spring from it—back to a single idea. This idea, called the "notion of limit," appears in Figure 17 as the bedrock foundation upon which the calculus arch stands. And, if calculus stands on this single idea, then so does all of mathematics to which it leads. In a sense, the complete branch of mathematics known as "analysis" comes from this fundamental idea.

Calculus changed from a partially intuitive description of real-world problems into rigorous mathematics when the formal definition of the limit concept was set down by Weierstrass around 1850. The exact definition is technical and I will not attempt to describe it here. I want to point out only that it exists and that it underlies much of mathematics. I know of no other mathematical notion so fruitful. Nor of any other which so clearly belongs in Bertrand Russell's collection of "great things."

The precise definition of limit involves inequalities, absolute values, two logical quantifiers, and a single implication—the combination of which accounts for its complexity. But each of these, taken individually, is quite simple and was considered back in my school days when mathematics was being told to me.

Before my teachers told me mathematics, they told me of its practical value. By "value," they meant that mathematics would help me get through each day. They meant it would help me with the "fact of living." They meant only this and nothing more.

They were misinformed.

I wish they'd meant by value that mathematics would take me here: to the base of the calculus arch, to the limit. Then, they would have been right.

PROMISES

In the beginning, I promised ideas.

"I'm going out for a stroll," I said. "Come along. We'll fetch some."

Let's see what we've found.

Our earlier discussion of the development of the number system follows standard lines and I claim no originality. My intention was to illustrate partially the fact (surprising to nonmathematicians) that all of mathematics flows from a few fundamental principles. Many details are omitted from the development but, for my purpose, they are unimportant. I wanted only to indicate that one could start with just the counting numbers 1, 2, 3, . . . , and five assumptions (the Peano Axioms), and then systematically create the integers, the rationals, the reals, and the complex numbers.

Once the real and complex numbers had been set down, the raw material of mathematics was on hand. I then turned to the relationship between pure and applied mathematics and how they come together in the applications process of Figure 7. This process has been discussed by others but not, I think, in exactly the way it is presented here. Figure 7—and the accompanying discussion—indicates clearly that the analysis stage takes place entirely in the mathematical world, while the application stage provides the connection between this world of ideas and the real-world of ordinary physical phenomena. This part of the process has about it such an air of magic and mystery that a distinguished physicist was led to speak of the "unreasonable effectiveness" of mathematics.

One of the by-products of the discussion associated with Figure 7 is the formulation of a clear distinction between pure and applied mathematics; namely, pure mathematics is analysis in the mathematical world; applied mathematics is pure mathematics which has a "pre-image" in the real world. This distinction, so far as I am aware, has not been made before.

After we passed through this part of the book, our stroll took us, more or less, through aesthetic country. From this point on, the discussion centered on the world of aesthetics and how mathematics fits into this world. As I see it, we fetched here four ideas which have some claim to originality and which are worth recounting. Let's list them now as a kind of summing up.

The ideas are:

1. the concept of the aesthetic ring,
2. the notion of the Mathworld,
3. the principle of minimal completeness and the principle of maximal applicability, and
4. the type M culture and the type N culture.

The *aesthetic ring* has as its inspiration the early notion of psychical distance, due to Edward Bullough. However, Bullough's notion was less quantitative than mine, and he did not speak of "rings." The idea here is that the aesthetic reaction of an observer to an art object falls into three reasonably distinct categories: (i) the observer experiences the object aesthetically, (ii) the object fails to provide an aesthetic experience because the observer is too close (in the sense of aesthetic distance), or (iii) the object fails to provide an aesthetic experience because it is too far away. The aesthetic ring of the art object is the annulus associated with condition (i) and is shown in Figure 12. The area inside the inner boundary of the ring is the "underdistanced" area of condition (ii) and the area beyond the outer boundary is the "overdistanced" area of condition (iii).

These ideas turned out to be particularly applicable to mathematics when the subject is examined from the point of view of mathematicians, namely, as being itself an art object (see Figure 13).

The Mathworld arose from the observation that a recent institutional aesthetic theory (due primarily to the contemporary philosophers George Dickie and Arthur Danto) applies almost unchanged to a similar complex composed of mathematicians and mathematics artifacts. The Dickie–Danto complex is called the Artworld. The Mathworld is defined in such a way as to become an isomorphic copy of a subset of the Artworld. This means that, except for notation, the Mathworld becomes a part of the Artworld.

Thus, the developed aesthetic theory of the Artworld carries over to the Mathworld and, maybe for the first time, there exists a legitimate aesthetic theory which includes mathematics as art.

The aesthetic theory of the Mathworld provides a means for determining which mathematics artifacts are, and which are not, works of art. But the theory does not provide a mechanism for distinguishing between good art and bad art. To help with this determination, the *principle of minimal completeness* and the *principle of maximal applicability* were formulated. Each mathematical artifact can, theoretically, be evaluated in terms of these principles. An artifact which satisfies both principles is, by definition, accepted as a "good" work of art and is awarded the formal appellation "elegant."

Consequently, the mathematicians' intuitive notion of elegance now takes on a semiformal character in terms of a genuine aesthetic theory. The degree of formality of the notion of elegance of a mathematical artifact depends no longer on intuition or convention, but rather on the degree of precision with which the two principles may be applied to it.

Finally, our stroll took us into the two cultures of C. P. Snow. We observed that the accepted characterization of these two groups (as scientists and humanists, respectively) is less precise than we would like. A sharper division is obtained by considering the cultures of *type M* and of *type N*—the type M culture consisting of those people who have developed a certain level of mathematical skill and the type N culture being composed of those who have not.

Although these new groups agree closely with Snow's original notions, they have the conceptual advantage of being more clearly defined. As Figure 16 shows, they are determined precisely by the aesthetic ring of mathematics. The type M members are underdistanced from mathematics, in the sense of aesthetic distance, while the type N members are too far away. Moreover, this new conception provides a valuable prescription for positive change. The type M and type N cultures are separated by the aesthetic ring of mathematics. The way to bring them together is clear: widen the ring.

At present, the mathematicians solely inhabit the aesthetic ring. They live here because—for whatever reason—they see mathematics as art. Widening the ring requires that others come to see mathematics the same way. This, in turn, requires a redesigning of our mathematics education system so that this new vision becomes possible. It's that simple. And that complicated.

The summary ends our tour. We've come far. We've walked the real line and wandered the complex plane. We saw the glow of rings. We felt the clash of cultures. Numbers moved in the grass. Abstractions wafted past like mist. The promise of high art broke through like sunlight.

We've strolled far enough. We have, I think, fetched some ideas. But it's not for me to say. You decide.

CHAPTER 9

Epilogue

That summer in Vermont I read Robert Frost under a shade tree. I liked best the old poems—"Mowing," "Birches," "Dust of Snow"—the ones he wrote as he passed forty, which was then my age as I read him under my tree.

It was hard then—and now—to imagine Frost so young. Always he lives in my mind as I first saw him around 1957, moving from side to side behind a lecturn with a rolling gait, tweed jacket unbuttoned over a sweater, his ancient head white as his shirt, "saying" the old poems straight, without comment, and entirely from memory.

He did "Reluctance," the final poem in his first book, *A Boy's Will*. When he said the last six lines he looked—I swear—straight at me with those old, sparkling eyes:

> Ah, when to the heart of man
> Was it ever less than a treason
> To go with the drift of things,
> To yield with a grace to reason,

And bow and accept the end
Of a love or season?

"Ahhh," he said, looking at me, and you could hear the
long "h" on the end of the word, and hear it again three
words later in "heart." The old man snapped the "t" at the
end of "heart" the way his knees did then and mine do now
when I straighten.

"You have my attention, Mr. Frost," I told him si-
lently. "No going with the drift of things for me."

He caught my attention first all those years ago. He
caught it again that summer in Vermont, in the bicenten-
nial year of 1976.

My shade tree was behind the house. Farther behind
leaned a pasture hill sloping up toward taller trees. From
the hill you could look north through a notch in the Green
Mountains and see Canada if you knew just where Vermont ends and Canada begins.

East lay the New Hampshire border and, beyond that,
the villages of Derry and Franconia where Frost wrote
many of the *Boy's Will* poems. Later he would live in Vermont. But those early poems were written over there, east
of me, in New Hampshire.

I imagined him then sitting under a tree, writing the
poems I would read sixty years later. From time to time he
would look up, I told myself, and rest his eyes on Vermont's
green hills. I turned my chair east to meet his glance.

"We are working together," I told him from the heart,
"you write and I read."

To write poetry, Frost had to avoid the drift of things
around him. The Yankee farmers planted and plowed,
milked and mowed, going each day about the routine tasks
common to those everywhere who wrest their living from

land. As farmers drifted about him, Frost sat still. Sometimes the farmers would see him walking, looking at the sky and the trees, his face concentrated and strange. But mostly when they caught sight of him, he was sitting. If they looked carefully they might see a pencil move in his hand.

Ignoring the drift of things, Robert Frost sat still and wrote his deceptively simple poetry. He built poems out of old-fashioned rhyme, syntax, and linearity. He crafted them so they sounded like language and so they yielded a logical, patterned mode of thought. On the surface, the poems said something you wanted to hear. Underneath, like muscle, Frost layered complexity and depth.

Frost's poems stay with you because of their simplicity; you come back to them for their depth.

Many came to Frost's poems. Almost single-handedly, he turned a practical-minded nation toward poetry. To be sure, the turning was short lived. But it happened, and some of us remember the old man on national television, squinting against bright sunlight, saying "The Gift Outright," at the 1961 presidential inauguration of John F. Kennedy. Never before—or since—has serious poetry been so auspiciously presented in the United States. For a time, through his books, his reading, and his personality, Robert Frost brought poetry to multitudes.

For me—in Vermont and now—the drift of things had to do with mathematics. By 1976, Vietnam, the social revolution, and forces I cannot explain had dramatically altered the face of the American academy. The golden age of mathematics that began with *Sputnik* had ended. Not again in my lifetime, I was certain, would support and enthusiasm for mathematics research reach the high level it attained in the early sixties when I began my career as an assistant professor of mathematics. In the summer of 1976, the academy in general, and research mathematics in particular, lay in disarray. I knew then that the remainder of the twentieth century would not be kind to university professors.

But the uses of adversity can be sweet. I sat that summer under my shade tree and wondered if I might not find opportunity in this adversity. Since financial support for research is fading, I thought, and since there is growing discontent with the university, perhaps this is the propitious moment to bring meaningful change to college teaching. Perhaps misfortune in research will cause the mathematicians to give additional attention to teaching. The drift began when *Sputnik* brought the research money. Now, the money has slowed. So, let's do something serious about teaching. Let's now chuck out the large classes, the trivial applications, and the cinder-block textbooks.

Yes, I thought, now it's time to ease aesthetics into the teaching of mathematics. Robert Frost brought poetry to the multitudes. Let's now bring them mathematics.

All this came to me that summer in Vermont. It came with the poetry. It came with the breeze and the sailing clouds. It came like a revelation.

I looked east again. "You and me together," I said.

Time passes. Fifteen years later, nothing has changed for the better. Classes are larger, textbooks clunkier, and the alleged applications even more trivial and more prevalent. The inability of American students to compete in international mathematics has become apparent. The culture of research mathematics remains invisible, unseen now even by undergraduates who major in the subject. The two cultures fly apart like galaxies on opposite sides of an expanding universe.

My Vermont revelation led to no revolution. Mathematics has not found its Robert Frost.

Maybe this is not a game for poets exactly as war is not

a game for knights. In order to teach mathematics properly, you surely need a poet. But to get him inside the classroom you may need something else. There is likely to be a fight at the schoolhouse door. And, like Horace, all poets may write better than they fight. To get inside, you may need a hooligan.

The revolution in mathematics teaching will come as certainly as sunrise. But it may be a revolution followed— not preceded—by poetry.

At any rate, I am not Robert Frost, nor was meant to be. I see myself now—somewhat romantically—more like an aging gunfighter, walking alone the mean streets of academe. But I can no more stop the academic drift of things than a turn-of-the-century shootist could stop the changing times and the settling in of law and order. In a while, the profession will just pass me by.

The gunfighter always found another town to tame. So far, I have always found another course to teach. But the Vermont summer passed long ago. Courses are running out. Soon I will be down to just one.

Just one more course and I'm done. Make it classical complex variables. Let me do it once more.

One day when the wind is right I'll do the Cauchy Integral Formula for the last time and I will do it truly. I will write it carefully and the students will see the curve and the thing inside and the lazy integral which makes the function value appear as quickly as my palm when I open my hand.

They will see the art of mathematics. And they will never care for anything half as much.

Notes

INTRODUCTION

1. G. H. Hardy, *A Mathematician's Apology* (London: Cambridge, 1973), p. 61.
2. Bertrand Russell, *Mysticism and Logic* (New York: Doubleday, 1917), p. 57.
3. Ian Stewart, *The Problems of Mathematics* (New York: Oxford, 1987).
4. Robert Frost, *The Poetry of Robert Frost* (New York: Holt, Rinehart & Winston, 1969), p. 1.

CHAPTER 1

1. James Dickey, *Poems 1957–67* (New York: Collier, 1968).
2. W. B. Yeats, *Selected Poems* (New York: Collier, 1962).

CHAPTER 2

1. John 18: 37–38.
2. Alfred Renyi, *Dialogues on Mathematics* (San Francisco: Holden-Day, 1967), p. 11.

3. Josephine Tey, *The Daughter of Time* (New York: Berkley, 1959), p. 122.
4. William Shakespeare, *The Tragedy of King Richard III, act 4*, sc. 3.
5. James R. Newman, *The World of Mathematics* (New York: Simon & Schuster, 1956), p. 728.
6. Jacob Bronowski, *Science and Human Values* (New York: Julian Messner, 1956), pp. 58–60.
7. James G. Frazer, *The Golden Bough* (Toronto: Macmillan, 1950), pp. 825–826.
8. Stewart, *The Problems of Mathematics*, p. 150.
9. Alfred Adler, "Mathematics and Creativity," *New Yorker*, Feb. 19, 1972, pp. 39–45.
10. Hardy, *A Mathematician's Apology*, p. 70.
11. Russell, *Mysticism and Logic*, p. 56.
12. Morris Kline, *Mathematics in Western Culture* (New York: Oxford, 1959), p. 428.
13. David Billington, *The Tower and the Bridge* (Princeton, NJ: Princeton University Press, 1985), pp. 9, 15.

CHAPTER 3

1. Quoted in David Burton, *Elementary Number Theory* (Boston: Allyn & Bacon, 1976), p. v.
2. Russell, *Mysticism and Logic*, p. 64.
3. Kurt Gödel, *On Formally Undesirable Propositions of Principia Mathematica and Related Systems* (New York: Basic, 1965).
4. Michael Guillen, *Bridges to Infinity* (New York: Tarcher, 1983), p. 20.
5. Morris Kline, *Mathematics, the Loss of Certainty* (New York: Oxford, 1980), p. 263.
6. Stewart, *The Problems of Mathematics*, p. 214.
7. Guillen, *Bridges to Infinity*, p. 125.
8. Russell, *Mysticism and Logic*, p. 70.
9. Albert Einstein, "Geometry and Experience," in Readings in the Philosophy of Science, eds. Herbert Feigl and May Brodbeck (New York: Appleton-Century-Crofts), p. 189.
10. Salomon Bochner, *The Role of Mathematics in the Rise of Science* (Princeton, NJ: Princeton University Press, 1966), p. 48.
11. Guillen, *Bridges to Infinity*, p. 64.
12. Ibid.
13. Thomas Tymoczko, "The Four Color Problems," *Journal of Philosophy, 76* (1979), pp. 51–53.
14. Stewart, *The Problems of Mathematics*, p. 117.

15. Kenneth Appel and Wolfgang Haken, "The Four-Color Problem," in *Mathematics Today,* ed. Lynn Steen (New York: Springer Verlag, 1980).

CHAPTER 4

1. Alfred North Whitehead, *Science and the Modern World* (New York: Macmillan, 1925), p. 34.
2. Eugene P. Wigner, "The Unreasonable Effectiveness of Mathematics in the Natural Sciences," *Communications on Pure and Applied Mathematics, 13* (1980), pp. 1–14.
3. "Mathematical Sciences: A Unifying and Dynamical Resource, Report of the Panel on Mathematical Sciences," *Notices of the American Mathematical Society,* June 1986, p. 465.
4. Ibid.
5. "Mathematics, the Unifying Thread in Science," *Notices of the American Mathematical Society,* October 1986, p. 725.
6. Bochner, *The Role of Mathematics,* p. 111.
7. Ibid., p. 190.
8. Wigner, "The Unreasonable Effectiveness," p. 14.

CHAPTER 5

1. Lynn Steen, ed., *Mathematics Today,* p. 10.
2. Thomas Munro, *Toward Science in Aesthetics* (New York: Liberal Arts Press, 1956), p. 3.
3. Arthur Berger, P. W. Prall, in *Aesthetic Analysis* (New York: Crowell, 1936), p. ix.
4. Nicholas Wolterstorff, *Works and Worlds of Art* (Oxford: Clarendon, 1980), p. v.
5. John Keats, "Ode on a Grecian Urn."
6. Henri Poincaré, *The Foundations of Science* (New York: Science Press, 1929), p. 386.
7. Ibid., p. 385.
8. Ibid., p. 391.
9. Seymore A. Papert, "The Mathematical Unconscious," in *On Aesthetics and Science,* ed. Judith Wechsler (Boston: Birkhauser, 1988), p. 106.
10. G. H. Hardy, *A Mathematician's Apology,* p. 92.
11. Ibid., pp. 85, 90.

12. Papert, "The Mathematical Unconscious," p. 106.
13. Ibid., p. 111.
14. Ibid.
15. Ibid., p. 112.
16. Ibid., p. 118.
17. Herman Weyl, "Symmetry," in *The World of Mathematics,* Vol. 1, ed. James R. Newman (New York: Simon & Schuster, 1956).
18. James Jeans, "Mathematics and Music," in *The World of Mathematics,* Vol. 4, ed. James R. Newman (New York: Simon & Schuster, 1956).
19. Gustav Fechner, *Vorschule der Aesthetik.* (Leipzig: Breitkoph & Hartel, 1876).
20. I. C. McManus, D. Edmondson, and J. Rodger, "Balance in Pictures," *British Journal of Psychology, 76* (1985), pp. 311–324.
21. George David Birkhoff, "Mathematics of Aesthetics," in *The World of Mathematics,* Vol. 4, ed. James R. Newman (New York: Simon & Schuster, 1956).
22. George Stiny and James Gips, *Algorithmic Aesthetics* (Berkeley: University of California Press, 1978).
23. Mortimer Adler, *Six Great Ideas* (New York: Macmillan, 1981).
24. George Dickie, *The Art Circle* (New York: Haven, 1984).
25. Will Durant, *The Story of Philosophy* (New York: Pocket Library, 1954), p. xxvii.
26. Allan Bloom, *The Closing of the American Mind* (New York: Simon & Schuster, 1987), p. 271.
27. Norman Bryson, *Vision and Painting* (London: Yale University Press, 1983).
28. Roger Scruton, "Recent Aesthetics in England and America," in *Aesthetics and Art Education,* eds. Ralph A. Smith and Alan Simpson (Chicago: University of Illinois Press, 1991), p. 43.
29. Arthur Danto, *Connections to the World* (New York: Harper & Row, 1990).
30. Harold Bloom, as quoted in L. S. Klepp, "Everyman a Philosopher," *New York Times* Magazine, Dec. 2, 1990, p. 117.
31. Arthur Danto, "The Artworld," *The Journal of Philosophy, 61* (1964), pp. 571–584, p. 572.
32. Ibid., p. 573.
33. Ibid., p. 581.
34. Eugene F. Kaelin, "Why Teach Art in the Public Schools?" in *Aesthetics and Art Education,* eds. Ralph A. Smith and Alan Simpson (Chicago: University of Chicago Press), p. 165.
35. Danto, "The Artworld," p. 571.
36. Dickie, *The Art Circle,* p. 11.
37. Ibid., p. 66.
38. Ibid., p. 67.
39. Ibid., p. 84.
40. Ibid., p. 80.

41. Paul Halmos, as quoted in *Mathematics Today,* Steen, ed., p. 1.
42. Richard Wollheim, *Art and Its Objects* (New York: Harper & Row, 1968).
43. R. G. Collingwood, *The Principles of Art* (London: Clarendon, 1938), p. 130.
44. Hardy, *A Mathematician's Apology,* p. 61.
45. Howard Nemerov, as quoted in Doug Anderson, "Poet in Prose," *New York Times Book Review,* April 28, 1991, p. 15.
46. David Hilbert, as quoted in Morris Kline, *Mathematics in Western Culture* (New York: Oxford, 1953), p. 397.
47. Poincaré, as quoted in Kline, *Mathematics in Western Culture,* p. 397.
48. C. P. Snow, as quoted in Hardy, *A Mathematician's Apology,* p. 37.
49. Edward Bullough, "Psychical Distance as a Factor in Art and an Aesthetic Principle," *British Journal of Psychology,* 5 (1912), p. 87–118.
50. Donald Sherburne, *A Whiteheadean Aesthetic* (New York: Yale, 1961), p. 108.
51. James L. Jarrett, *The Quest for Beauty* (Englewood Cliffs, NJ: Prentice Hall, 1957), p. 111.
52. Bullough, "Psychical Distance," p. 92.
53. Ibid., p. 93.
54. Ibid., p. 94.
55. Ibid., p. 95.
56. Ernest Hemingway, *Death in the Afternoon* (New York: Scribners).
57. Ortega y Gasset, *The Dehumanization of Art and Notes on the Novel* (Princeton, NJ: Princeton University Press, 1948), p. 12.
58. Ernest Nagel and James R. Newman, *Gödel's Proof* (New York: NYU Press, 1958), p. 24.
59. Martin Gardner, *Mathematical Puzzles and Diversions* (New York: Simon & Schuster, 1959), p. 47.
60. Angus E. Taylor, *Advanced Calculus* (Boston: Ginn, 1955).
61. Martin Gardner, *2nd Book of Mathematical Puzzles and Diversions* (New York: Simon & Schuster, 1961), p. 57.
62. George B. Thomas, *Calculus and Analytic Geometry* (Reading, MA: Addison-Wesley, 1951).

CHAPTER 6

1. Bloom, *The Closing of the American Mind,* p. 25.
2. Hazzard Adams, *Academic Tribes* (New York: Liveright, 1976), p. 8.
3. Morris Kline, *Why the Professor Can't Teach* (New York: St. Martin's, 1977), p. 240.
4. P. J. Hilton, "Teaching and Research: A False Dichotomy," *The Mathematical Intelligencer, 1,* (1978), pp. 76–80.

5. Timothy O'Meara, "Strategies for Enhancing Resources in Mathematics," *Notices of the American Mathematical Society, 33,* (1986) p. 327.
6. Kline, as quoted in, "Mathematics and the Dilemma of University Education," *The Mathematical Intelligencer, 1* (1978), p. 10.
7. John Barton, *Playing Shakespeare* (London: Methuen, 1984), p. 50.
8. O'Meara, "Strategies," p. 329.
9. Ibid., p. 327.
10. Adler, "Mathematics and Creativity," pp. 39–45.
11. L. T. More, *The Limitations of Science* (New York: Holt, 1915), p. 150.
12. Ibid., p. 151.
13. Adler, "Mathematics and Creativity," p. 41.
14. Ibid., p. 45.
15. Ibid., p. 41.
16. Ibid.
17. Edna St. Vincent Millay, *Collected Sonnets* (New York: Harper, 1988), p. 45.
18. William Shakespeare, *Troilus and Cressida,* act 1, sc. 1.
19. Bertrand Russell, *In Praise of Idleness* (New York: Unwin, 1962), p. 127.
20. William Shakespeare, *Henry IV,* Part I, act 1, sc. 1.
21. Adler, "Mathematics and Creativity," p. 41.
22. Ibid.
23. Ronald C. Douglas, "Castle in the Sand," *Calculus for a New Century* (Mathematical Association of America, 1988).
24. Gina Bari Kolata, "Calculus Reform: Is it Needed? Is it Possible?" *Calculus for a New Century,* Lynn Steen, Ed. p. 89.
25. Richard W. Hamming, review of *Toward a Lean and Lively Calculus, American Mathematical Monthly,* May 1988, pp. 466–471.
26. Leonard Gillman, "The College Teaching Scandal," *Focus 8*, (1988), p. 5.
27. R. Emmett Tyrrell, Jr., "A Conservative Crack-up?" *Wall Street Journal,* March 27, 1987.
28. More, *The Limitations of Science,* p. 150.
29. Russell, *Mysticism and Logic,* p. 57.

CHAPTER 7

1. C. P. Snow, *The Two Cultures and a Second Look* (New York: Cambridge, 1964), p. 3.
2. Ibid., p. 62.
3. Ibid., p. 63.
4. Ibid., p. 9.
5. Jacques Barzun, *A Stroll with William James* (New York: Harper & Row, 1983), p. 203.

6. Ibid., p. 206.

7. Bloom, *The Closing of the American Mind,* pp. 182, 350.

8. Snow, *Two Cultures,* p. 14.

9. Ibid., p. 13.

10. Stewart, *The Problems of Mathematics.*

11. Augustus De Morgan, "Assorted Paradoxes" in *The World of Mathematics,* ed. James R. Newman (New York: Simon & Schuster, 1956), p. 2378.

12. Hedley Bull, "Tradition and Science in the Study of International Politics," in *Contending Approaches to International Politics,* Klaus Knorr & James N. Rosenau, eds. (Princeton, NJ: Princeton University Press, 1969), pp. 26–30.

13. Morton A. Kaplan, "The New Great Debates," in *Contending Approaches to International Politics,* Knorr and Rosenau, eds., p. 55.

14. Ibid., p. 61.

15. Allen L. Hammond, "Mathematics, Our Invisible Culture," in *Mathematics Today,* ed. Lynn Steen (New York: Springer Verlag, 1980), p. 16.

16. Ibid., p. 16.

17. John Polkinghorne, *One World* (Princeton, NJ: Princeton, 1986), p. 46.

18. Collingwood, *The Principles of Art,* p. 331.

19. Ibid., p. 312.

20. Ibid.

CHAPTER 8

1. Russell, *Mysticism and Logic,* p. 55.

2. Robert Henri, *The Art Spirit* (New York: Harper & Row, 1951), p. 265.

Index

About the Author

Jerry P. King, Ph.D., is Professor of Mathematics at Lehigh University. He served as Dean of the Lehigh Graduate School in 1981–1987 and as Associate Dean of Arts and Sciences in 1979–1981. In 1986, King was chosen for the Deming Lewis Faculty Award of the Lehigh Alumni Association.

King served as President of the Pennsylvania Association of Graduate Schools and as a member of the Board of Governors of the Mathematical Association of America. His interests include music, poetry, art, and athletics.

Dr. King has published many mathematics research papers, mainly in the areas of complex analysis and summability theory. Well known as a speaker, King has lectured on mathematics nationwide.

He received his bachelor's degree in electrical engineering from the University of Kentucky in 1959. In 1962, he received the doctor of philosophy in mathematics from the same institution.